Bears & Other Playthings

Past & Present

DEE HOCKENBERRY

PHOTOGRAPHY BY TOM HOCKENBERRY

4880 Lower Valley Road, Atglen, PA 19310 USA

DEDICATION

For Tom, without whose help
I could not accomplish anything.

Hockenberry, Dee.
Steiff bears & other playthings past & present / Dee Hockenberry ; photography by Tom Hockenberry.
p. cm.
Includes bibliographical references and index.
ISBN 0-7643-1120-4 (hb)
1. Margarete Steiff GmbH. 2. Toys--Germany--History--19th century. 3. Toys--Germany--History--20th century. I. Title: Steiff bears and other playthings past & present. II. Hockenberry, Tom. III. Title.

NK9509.65.G34 M3734 2000
688.7'26--dc21

99-087327

Book Design by Anne Davidsen
Type set in Lithograph / Humanist 521

ISBN: 0-7643-1120-4
Printed in China
1 2 3 4

Published by Schiffer Publishing Ltd.
4880 Lower Valley Road
Atglen, PA 19310
Phone: (610) 593-1777; Fax: (610) 593-2002
E-mail: Schifferbk@aol.com
Please visit our web site catalog at
www.schifferbooks.com or write for a free catalog.
This book may be purchased from the publisher.
Please include $3.95 for shipping.

In Europe, Schiffer books are distributed by
Bushwood Books
6 Marksbury Ave.
Kew Gardens
Surrey TW9 4JF England
Phone: 44 (0)208 392-8585
Fax: 44 (0)208 392-9876
E-mail: Bushwd@aol.com
Free postage in the UK. Europe: air mail at cost

Please try your bookstore first.

We are interested in hearing from authors
with book ideas on related subjects.

CONTENTS

ACKNOWLEDGMENTS

It is my extreme pleasure and my immense good fortune to have a lovely relationship with the Steiff Company. I count Jörg Juninger, Susanna Pinyuh, and Barbara Ehrlinger (all members of the Steiff family) as friends of many years standing. Paul Johnson, Barbara Sprenger and Rita Bagala, of Steiff U. S. A., have always been supportive of my work. I am indebted to them all and extend my heartfelt thanks as well as a special bear hug to Susanna for her wonderful preface.

The author with Susanna Steiff Pinyuh (left) and Barbara Steiff Ehrlinger at Barbara's house in Giengen, Germany. Susanna and Barbara are Margarete Steiff's great grand nieces and Richard Steiff' s granddaughters.

The author with (from left to right) Paul Johnson, president of Steiff U S A, Tweed Roosevelt, Teddy Roosevelt's great grandson and Jörg Juninger, Margarete Steiff's great grand nephew. Tweed represents the company at events that celebrate the Teddy Bear. Jörg has been an officer in the company since 1970, serving as the manager of development and design. He is the director of the museum and is in charge of the archives. Festival of Steiff 1999.

PREFACE

By Susanna Steiff Pinyuh

I don't know if it is possible to adequately describe the importance of Aunt Margarete to the firm. My mother always said that Margarete was the driving force during the years she was alive and vigorous, but her influence lasted for years, even after her death. Indeed, to this day, people in Giengen are still talking about her and other family members, and what an unforgettable experience it was to work at the Steiff Company then! Her motto "nothing is too good for the children" continues to be strongly upheld.

In a long letter written in the spring of 1903 by Margarete's niece Lina to Lina's brother Paul, who had been sent on a long trip abroad on Steiff company business, Lina writes: "Our good Aunt (Margarete) with her great ambition and fine health, is at the helm day and night, now upstairs, now downstairs, always energetic and excited, always throwing herself into the work, including all the aggravation that comes with it. She expects every little detail to be taken care of immediately." "(Margarete was known as a tough taskmaster.)"

Naturally, we are all extremely proud of our amazing Tante Margarete's achievements. She created a company which not only produced a charming multiplicity of toys, but forged important and lasting business connections the world over. And she managed all this before the turn of the 20th century—an era when women certainly were not expected to achieve business success.

So much has been written about the Steiff men and their very considerable contributions, but it really should be noted that many of the other women in the family, from my great grand aunt Margarete's nieces on down to their daughters and granddaughters, have been extremely useful to the firm in one productive capacity or another.

It is so gratifying to me personally, and to all the Steiff family in the United States and Germany, that Dee, herself a very ambitious and hard working woman, has produced another intriguing book, this time chronicling the evolution of Steiff Toys. I know you will feel very rewarded no matter how often you consult this fine work.

INTRODUCTION

STEIFF PLAYTHINGS, "MUSIC" FOR THE SOUL

Webster's dictionary defines "Music Of The Sphere's" as an ethereal music supposed, by early mathematicians, to be produced by the movements of the heavenly bodies. We, who love Steiff playthings, think of our collections as music for our souls and heavenly indeed. Of course, these wonders are not produced by the stars in the sky, but by artisans and craftsmen who, for over 100 years, have been stars of the toy world.

CHAPTER I

THE MUSIC BEGINS

THE WONDERFUL MUSIC OF STEIFF

The time—July 24, 1847. The place—Giengen en Brenz, Germany. The event—the birth of a third daughter to Maria and Friedrich Steiff. The child was christened Apollonia Margarete, but subsequently was known by her second name. No one could guess that this tiny morsel of humanity, and the ensuing events, would forever more impact the toy and collecting world.

At the tender age of eighteen months Margarete contracted polio, a dreaded and debilitating disease. Her parents tried everything available to procure a full recovery, but their efforts were futile. The child was destined to spend the rest of her life in a wheelchair for her left foot was totally paralyzed, her right foot was lame and she had only partial use of her right arm. Perhaps a lighter spirit would bemoan the slings and arrows that fate had bestowed, but Margarete proved to have strength, courage and a will to become far more than most women of her generation.

Margarete Steiff attended school and was even carried, by stronger children, to some of her classes. She learned to sew on a machine and, though painful, became proficient in hand needlework as well. With the completion of her education she was employed as a dressmaker and succeeded not only because of her skill, but as a result of her vibrant personality. She was always a joy to be associated with.

In 1880, she made her first toy fashioned of felt and stuffed with lambs wool. The tiny elephant delighted the children who received it; as it did the adults who used it as a pin cushion. She even sold a few (although production was limited, reaching only 29 by 1882.) Soon new animals were created. Ten years later, this extraordinary woman had formed a company and applied for patents to make other playthings.

Margarete never married, but her siblings did and, to this day, their progeny carries on what she began. It was her brother Fritz who was the chief instrument in helping her company gain a foothold. He realized his plans, with the aid of five of his six sons, and soon the Steiff organization became a legitimate and successful enterprise. On March 3, 1893 the business was registered at the Chamber of Commerce and employed fourteen workers, ten of whom worked at home. The rest, as the saying goes, is history.

Her nephew, Richard, one of Fritz sons, was an artist of remarkable ability and a driving force behind the creative designs. He often sketched animals at the zoo and in 1902 was inspired to devise a bear with articulated limbs and head. This toy was listed in the firm's records and premiered at the Leipzig Toy Fair in 1903. "Baerle" was string jointed through inner discs and, although many were made and offered, none seemed to have survived. The next model was jointed by metal rods and, while extremely rare, they do surface on occasion.

So begins the saga of the Teddy Bear, a phenomenal soft toy named after a President of the United States, Teddy Roosevelt. While teddies have remained the all time popular product in Steiff's line, the magnificent animals, dolls, and other wonders are adored by generations of children and adult collectors alike. So now it's show time. Let the music begin.

Bronze statue at the entrance to the Steiff Museum and factory in Giengen, Germany.

View of the factory in 1912.

Portrait of Margarete Steiff, honoring her 150th birthday in 1997.

Circa 1910, donkey and deer on metal wheels; circa 1925, Bully dog on wooden wheels. Steiff Museum.

Circa 1910, deer on metal wheels. Steiff Museum.

Circa 1910, pig, chicken, lamb, dog and two deer. Steiff Museum.

A variety of early monkeys and other animals. Steiff Museum.

Early dolls and goat. Steiff Museum.

Circa 1910, cow and goat on wheels. Steiff Museum.

Dwarves and dressed animals. Steiff Museum.

Circa 1900, Skittles and other animals posed in front of a shop display. Steiff Museum.

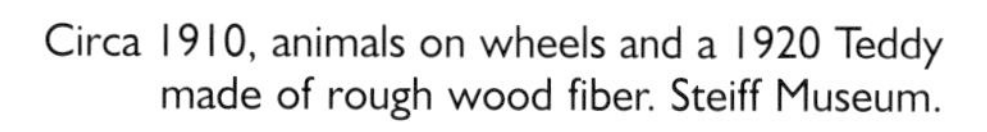

Circa 1910, animals on wheels and a 1920 Teddy made of rough wood fiber. Steiff Museum.

1912 Monkey radiator cap and other early animals. Steiff Museum.

1930 and 1940 animals. Steiff Museum.

1910 to 1930 dolls and animals. Steiff Museum.

1990 baby animals. Steiff Museum.

Jungle display. Steiff Museum.

Jungle display. Steiff Museum.

Mechanical Village. Steiff Museum.

1911 Clowns. Archive pieces at Steiff Festival in Toledo, Ohio.

1920 red elephant. Archive example at Steiff Festival in Toledo, Ohio.

1928 *Fluffy* cats. Archive examples at Steiff Festival in Toledo, Ohio.

1926 *Molly* dogs. Archive examples at Steiff Festival in Toledo, Ohio.

1926 *Teddy Rose*. Archive display at Steiff Festival in Toledo, Ohio.

1932 *Jumbo* elephants. Archive display at Steiff Festival in Toledo, Ohio.

1927 *Bonzo.* Archive display at Steiff Festival in Toledo, Ohio.

1955 Santa and Reindeer. Archive display at Steiff Festival in Toledo, Ohio.

1926 *Spark Plug* and *Barney Google.* Archive display at Steiff Festival in Toledo, Ohio.

1914 pig on wheels; 1916 calf on wheels; 1914 lamb; various 1950s examples. Archive display at the Margaret Strong Museum, Rochester, New York.

1930 to 1950 animals, including an oilcloth rabbit. Archive display at the Strong Museum, Rochester, New York.

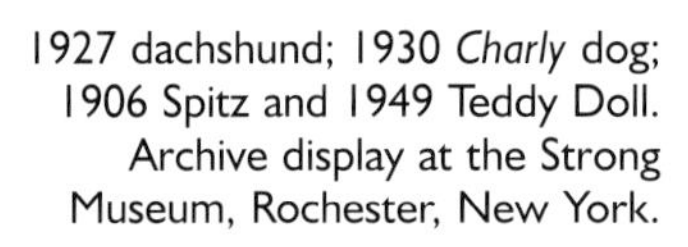

1927 dachshund; 1930 *Charly* dog; 1906 Spitz and 1949 Teddy Doll. Archive display at the Strong Museum, Rochester, New York.

1911 Felt Dolls. Archive display at the Strong Museum, Rochester, New York.

1913 Fireman and 1911 boy doll. Archive display at the Strong Museum, Rochester, New York.

1893 Bulldog. Archive display at the Strong Museum, Rochester, New York.

1920-1930 cats and dogs. Archive display at the Strong Museum, Rochester, New York.

1914 Rabbit and 1895 Stork. Archive display. Strong Museum, Rochester, New York.

1938 Koala Bear. Archive display. Strong Museum, Rochester, New York.

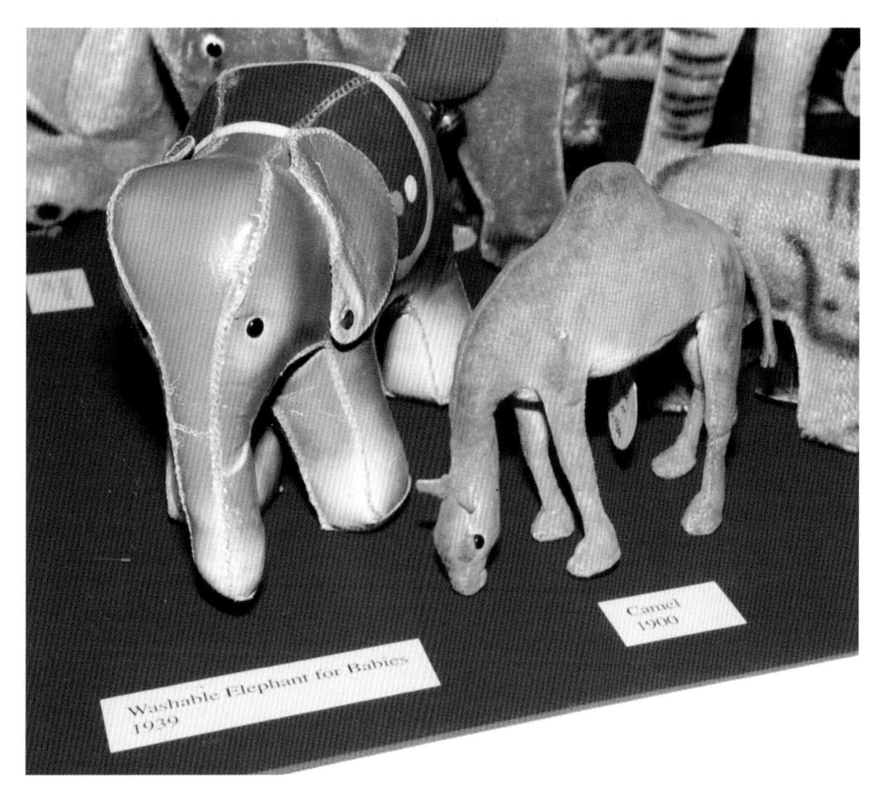

1939 washable elephant and 1900 camel. Archive display. Strong Museum, Rochester, New York.

CHAPTER II

TEDDY BEARS THEN & NOW

THE TEDDY BEARS NOT ONLY HAD A PICNIC BUT A HUNDRED YEAR OUTING.

Enter the Teddy Bear. In the toy community, dolls had reigned supreme for many years. Then a soft toy, fashioned in the image of a rather fierce creature, was introduced and ultimately became the most popular plaything of all time.

By 1902 the Steiff company had a full range of animals in mohair and felt. With the Leipzig Toy Fair fast approaching in the spring of 1903, it seemed imperative to present something new to compete with competitive companies' booming doll trade. Margarete's nephew, Richard, often sketched bears at the Stuttgart Zoo and it was ultimately decided that, since it was an animal of universal interest, this might be what they were looking for. A bruin was string jointed in the same manner as the dolls Richard had been experimenting with and was classified as 55 PB. Although it proved to be rather cumbersome, the sample was duly introduced at the Toy Fair. No one was remotely interested until at the last moment a buyer from the Geo. Borgfeldt Co. in New York City stopped to chat at the Steiff booth. When he saw the bear, the buyer became excited and promptly ordered 3,000.

Back in Giengen, it was back to the drawing board. A new design, called 34 PB, was ready for the 1904 World's Fair in St. Louis. This product was smaller, less bulky and of a more pleasing design. In fact, so pleasing that a fantastic number, totaling 12,000, were sold.

The string joining eventually proved unsatisfactory and ultimately metal rods, (discernible by X rays) replaced the stringing. Although completely efficient, the rods were found to be too expensive and were discontinued after a year. The last method tried was to employ round discs. This method was so successful that articulation by this means remains in use today. Several changes in design also took place and soon the conformation of the teddy bear, as we know it today, was accomplished.

Mohair, a natural fiber garnered from the angora goat, was and is the fabric of choice. During times of stress, such as both World Wars, it became scarce and other materials were substituted. All of the bears (and animals) in this book will be made of mohair unless otherwise noted. The teddies will also be fully jointed (head and four limbs), stuffed with excelsior, have embroidered features and claws, and felt will be used for the paw pads. Any variables will be commented upon in the photo captions. The differences in price range, for what may seem to be similar bears, is reflected by condition, rarity, color, and that indefinable quality called appeal.

Through the years the Steiff Company reached for the stars to bring the public new, exciting and innovative teddies. As the following pages unfold you will be able to delight in this fascinating evolution.

Rod Bear: 15 inches. Apricot mohair; sealing wax nose (replaced) tan twine mouth; five black claws; very hard stuffed body; joined by metal rods; elephant button. 1903. $25,000 up

Teddy: 15 inches. Cinnamon mohair; shoe button eyes; embroidered nose, mouth and five claws; this bear followed the rod bear; very softly stuffed excelsior; no I. D. 1903-1904. $6,000 up

Rod Bear. Showing unusual conformation.

The rod and the teddy that followed him, showing the different designs.

Center seam Teddy: 20 inch white mohair; shoe button eyes; pinkish tan embroidered nose, mouth and four claws; the head has a center seam (used by the company at the end of the bolt to conserve fabric and use it all); since there were fewer center seam bears it is eagerly sought; 1907; no I. D. $17,000

Center Seam Teddy: 24 inch apricot mohair; shoe button eyes; circa 1906/07. No I. D. $18,000 up

Teddy: 30 inches. Pale apricot mohair; felt pads; shoe button eyes; excelsior and soft stuffed; center head seam; circa 1906/1907; FF printed button. $17,000

Teddy Bear: 20 inches. Cinnamon gold mohair; shoe button eyes; circa 1907; blank button. $7,000 up

Teddy Bear: 16 inch. Pale cinnamon mohair; shoe button eyes; circa 1907; blank button. $4,000

Teddy Bear: 13 inch. Off white mohair; shoe button eyes; circa 1907; blank button. $3,500 up

Teddy: 12.50 inches. Off white mohair; shoe button eyes; circa 1907; blank button. $3,000 up

Teddy: 10 inches. Off white mohair; shoe button eyes; circa 1907; blank button. $2,200 up

Teddy: 10 inches. Worn mohair; shoe button eyes; patched pads; sweater not original; blank button; circa 1907. $800-$1,000

Teddy: 13 inch. Tan mohair; shoe button eyes; damage to muzzle area. Circa 1907. Blank button. $800-$1,000

Teddy: 14 inch. Bright cinnamon mohair; shoe button eyes; circa 1906; blank button. $2,900

Teddy: 13.50 inches. Gold mohair; shoe button eyes; circa 1906; blank button. $4,200

Cone Nose Teddy: 30 inches. Pale cinnamon mohair; shoe button eyes; circa 1907; no I. D. $20,000

Teddy: 24 inch. Creamy white mohair; shoe button eyes; rusty pink nose, mouth and claws; circa 1907/08. Printed FF button. $16,000

Teddy: 10 inches. Blond mohair; shoe button eyes; shows wear; circa 1907; blank button. $1,500-$2,000

Teddy: 13 inch. Pale cinnamon mohair; shoe button eyes; shows wear; circa 1909; no I. D. $1,700

Teddy: 14 inch. Gold mohair (thinning); shoe button eyes; circa 1908; printed FF button. $1,500-$1,600

Teddy: 20 inches. Apricot mohair; shoe button eyes; circa 1909; no I. D. $6,500 up

Teddy: 12 inch. Tan mohair; shoe button eyes; circa 1910; no I. D. $1,700
Teddy: 16 inch. Gold mohair; shoe button eyes; circa 1907; blank button; $5,500
Teddy: 14 inch. Gold mohair; glass eyes; circa 1920; printed FF button. $5,000
Dennis Yusa collection.

Teddy: 10 inches. Creamy white mohair; shoe button eyes; circa 1907. Blank button. $2,500

Teddy Bear: 10 inches. Tan mohair; shoe button eyes; wear to mohair; circa 1909; no I. D. $1,500

Teddy Bear: 10 inches. White mohair; shoe button eyes; circa 1908; Printed FF button. $1,700-$1,800

Teddy: 16 inch. Cinnamon mohair; shoe button eyes; circa 1907; blank button. $6.500
Teddy: 13 inch. Gold mohair; shoe button eyes; no I. D. Circa 1910. $3,000
Dennis Yusa collection.

Teddy Bear: 16 inch. Tan mohair; shoe button eyes; replaced felt pads; center head seam; worn condition; circa 1908. Printed FF button. $2,500 up

Teddy Bear: 14 inch. Pale silvery gold mohair; shoe button eyes; felt pads; circa 1907. No I. D. $4,000 up

Teddy Bear: 13 inch. Pale apricot mohair; shoe button eyes; excelsior and soft stuffing; circa 1907. No I. D. $3,500

Teddy Bear: 8 inch. Tan mohair; shoe button eyes; shows wear; circa 1908. Printed FF button. $850-$900

Teddy: 13 inch. Tan mohair; shoe button eyes; circa 1909; printed FF button. $3,900

Teddy Bear: 9 inch. Off-white mohair; shoe button eyes; felt pads (paws replaced); worn condition; circa 1907. Printed FF button. $1,000 up

Teddy: 24 inch. Cinnamon mohair; shoe button eyes; center head seam; circa 1907; blank button; $18,000

Teddy: 15 inches. Beige mohair; shoe button eyes; circa 1908; printed FF button. $4,500 up

Muzzle Bear: 18 inch. Shaded tan mohair; shoe button eyes; softly stuffed excelsior; original leather muzzle. No I. D. Circa 1909. $14,000

Teddy Girl: 18 inch. Cinnamon mohair; shoe button eyes; lifelong companion to Colonel T. R. Henderson, founder of Edinburgh, Scotland's, "Good Bears Of The World." Sold at Christie's auction in 1994 for approximately $172,000. Now resides in the Izu Museum, Japan.

Teddy: 20 inches. White mohair; shoe button eyes; suit was made for bear, but not issued with it. Circa 1907. Photo courtesy of Christie's. Sold at Christie's auction in London for $20,000 in 1995.

Muzzle Bear: 16 inch. Tan mohair; shoe button eyes; excelsior stuffed; original leather muzzle and leash; printed FF button; circa 1910. $10,000

Eliot: 13 inch. Blue mohair; shoe button eyes; excelsior stuffed; special order for Harrods in 1908. Photo courtesy of Christies. Sold at Christie's London in 1993 for $74,250.

Alfonzo: 12 inch. Red mohair; shoe button eyes; wearing a Cossack suit; this bear has a provenance of belonging to the Russian princess Xenia, relative of the Romanovs. 1908. Because of this and its one of a kind status, no accurate value can be placed on it. Photo courtesy of Ian Pout, owner.

Teddy Bear: 12 inch. Gold mohair; shoe button eyes; working squeaker; minor holes on pads; printed FF button; circa 1911. $1,600-$1,700

Teddy Bear: 13.50 inches. Off white mohair; patched and replaced pads; shoe button eyes; shows much wear; sweater not original; circa 1910; no I. D. $1,300

Hot Water Bottle Bear: 20 inches. Gold mohair; shoe button eyes; brass hooks and shoe lace open front; contains hot water bottle; 1907. Only 90 made from 1907 to 1914; printed FF button. Photo courtesy of Christie's, London. Sold at Christie's for $18,000.

Teddy Bear: 8 inch. Tan mohair; replaced felt pads; shoe button eyes; trim not original; circa 1910; no I. D. $500-$600

Teddy Bear: 20 inches. Gold mohair; shoe button eyes; shows wear; circa 1910; no I. D. $5,000

Teddy Bear: 13 inch. Cinnamon mohair; shoe button eyes; circa 1910; printed FF button. $3,900

Teddy Bear: 29 inch. Off white mohair; shoe button eyes; rusty pink embroidered nose, mouth and claws; felt pads patched; no I. D. $12,000 up

Teddy Bear: 13 inch. Faded cinnamon mohair; shoe button eyes; worn condition; circa 1910; no I. D. $1,500-$1,700

Teddy Bear: 16 inch. Cinnamon mohair; shoe button eyes; circa 1910; no I. D. $3,500 up

Teddy Bear: 18 inch. Dark apricot mohair; shoe button eyes; replaced pads; circa 1910; printed FF button. $4,000 up

Teddy Bear: 13 inch. Off white mohair; shoe button eyes; circa 1910 Shows some wear; printed FF button. $2,700

Teddy Bear: 14 inch. Off white mohair; shoe button eyes; excellent condition; circa 1910; no I. D. $3,900

Teddy: 12 inch. Gold mohair; felt pads w/moth holes; shoe button eyes; printed FF button; circa 1910. $1,800-$2,000

Teddy Bear: 12.50 inches. Off white mohair; shoe button eyes; circa 1910; printed FF button; trace of white stock tag. $3,900

Teddy: 9.50 inches. Tan mohair; shoe button eyes; shows wear; circa 1910; printed FF button. $1,500

Teddy: 20 inches. Short black mohair; shoe button eyes backed by red felt; no I. D. 1912. $22,000

Teddy: 12 inch. White mohair; glass eyes; flannel pads; printed FF button; circa 1912. $3,500

Bear Dolly: 12 inch. Red mohair body; white mohair head; white yarn ruff; shoe button eyes; printed FF button; circa 1913. Photo courtesy of Christie's. Sold by Christie's, London, in 1994 for $16,625.

Teddy: 20 inches. Curly black mohair; shoe button eyes; 1912. Photo courtesy of Christie's. Sold by Christie's London in 1994 for $35,000.

Teddy: 15 inches. Black mohair; black shoe button eyes backed by red felt; 1912. Photo courtesy of Christie's. Sold by Christie's in 1994 for $6,200 (a bargain. Resale value would be several times its cost).

Tumbling Teddy: 12 inch. Gold mohair showing wear; shoe button eyes; wind right arm to activate somersault action; circa 1910. $5,000 (a great deal more if mint or near mint).

Teddy: 20 inches. Gold mohair; shoe button eyes; shows wear; printed FF button; circa 1910 Lorraine Oakley collection. $2,500 up

Teddy: 9 inch. Cinnamon mohair; shoe button eyes; wearing old overalls of the period, but not original to the bear; no I. D. $1,900

Teddy: 13 inches. Off white mohair; shoe button eyes; shows some wear; no I. D; circa 1910. $1,700

Teddy: 16 inches. Tan mohair; shoe button eyes; circa 1910; printed FF button. $5,500

Teddy: 6 inches. Gold mohair; glass eyes; no pads; no I. D. Circa 1920 Hat not original. $1,000

Teddy 10 inches. Gold mohair; glass eyes; no I. D.; 1920s; dress not original. $1,600

Teddy: 10 inches. White mohair; glass eyes; no I. D.; 1920s. $2,800

Teddy: 12 inch. Gold mohair; glass eyes; printed FF button; 1920s. $3,000

Teddy: 12 inch. White mohair; glass eyes; printed FF button; 1920s. $3,200

Teddy: 13.50 inches. White mohair; glass eyes; linen pads; printed FF button; 1920s. $3,500

Teddy: 16 inches. White mohair; glass eyes; printed FF button; 1920s. $5,800

Teddy: 16 inches. Gold mohair; glass eyes; printed FF button; 1920s. $4,500

Teddy: 16 inches. Gold mohair; glass eyes; printed FF button; 1920s. $4,600. Clothing not original.

Teddy: 18 inches. Gold mohair; glass eyes; printed FF button; 1920s. $4,900

Teddy: 24 inches. Cream mohair with brown tipping; (faded) oversize glass eyes; 1928; printed FF button. $15,000 up

Teddy: 20 inches. White mohair; glass eyes; printed FF button with trace of white stock tag; superb example; 1920s. $14,000

Teddy: 25 inches. White mohair; glass eyes; printed FF button with trace of orange stock tag; 1928; superb example. $15,000. Clothes not original.

Teddy: 24 inches. Gold mohair; glass eyes; printed FF button; 1920s. $9,000

Teddy: 30 inches. Gold mohair; glass eyes; shown with 1925 photo of one of the original owners (it first belonged to her brother). Also included is a facsimile copy of a book written by her father. Printed FF button and trace of orange stock tag. 1920s. $16,000

Teddy Clown: 12 inches. Cream mohair (originally tipped brown;) glass eyes; original ruff; missing hat; printed FF button; 1926. Because of much wear, the price is devalued. $2,500 up

Teddy Clown: 12 inches. Cream mohair (originally tipped brown); glass eyes; original hat; pom poms and ruff replaced; 1926; no I. D. $5,500 up

Above:: *Petsy Bear*: 10 inches. Cinnamon tipped cream mohair; blue glass eyes; rosy red mouth, nose and claws; wired ears. No I. D. 1928. Too worn to evaluate.

Right: Teddy Clown: 19 inches. Beige tipped cream mohair; glass eyes; original hat and silk ruff; printed FF button; 1926. Photo courtesy of Christie's. Sold at Christie's, London, in 1996 for $12,000 (add value in today's market).

Petsy Bear: 24 inches. Cinnamon tipped cream mohair; oversize wired ears; blue glass eyes with milk glass on back; note that all Petsy's have a seam from ear to ear and from the center of this seam another seam runs down the center of face to the nose. Printed FF button; 1928. $15,000 up

Petsy Bear: 14 inches. White mohair; brown glass eyes; 1927; no I. D. Only example known, therefore cannot be evaluated.

Teddy: 15 inches. Brown mohair; glass eyes; printed FF button; 1930s. $5,900

Teddy Baby: 6 inches. Gold mohair; glass eyes; velvet muzzle and feet; open mouth; printed FF button; circa 1930. $2,000 up. Bow not original.

Dicky Bear: 13 inches. Gold mohair; inset muzzle; glass eyes; replaced pads; some wear; no I. D. Circa 1930. $1,800 up (a prime example would be many times that figure). Sweater added.

Teddy Baby: 11 inches. Tan mohair; clipped inset muzzle and feet; open mouth; original spiked collar; printed FF button; trace of orange stock tag; circa 1920. $3,000

Teddy Rose: 13 inches. Pink mohair (color restored). Glass eyes; printed FF button; circa 1930. $2,500 (original condition would exceed this valuation several times over).

Teddy: 12 inches. Gold mohair; glass eyes; printed FF button with trace of orange stock tag. Circa 1930. $3,900 up

Teddy: 24 inches. Gold mohair; glass eyes; printed FF button. Circa 1930. $9,000

Bear Doll: 9 inch. Rayon plush head and paws; rayon body and limbs; open mouth; glass eyes; printed button; circa 1945. $1,500 up

Teddyli Bear: 9 inch. Mohair head, top of paws and feet; open mouth; glass eyes; cotton body and shirt; felt jacket and pants; circa 1950; raised script button. $1,400 up

Teddy Baby: 3.50 inches. Brown mohair; velvet muzzle and feet; glass eyes; circa 1948. Square head bear chest tag. $1,300 up

Teddy Baby: 3.50 inches. Gold mohair; velvet muzzle and feet; glass eyes; square head bear chest tag; U S Zone tag; circa 1948. $1,300 up

Teddy Baby: 11 inches. Blonde curly mohair; clipped mohair muzzle and feet; open felt mouth; glass eyes; suedene paw pads (bottom cardboard lined to facilitate standing); leather collar; circa 1955. No I. D. $1,200 up

Music Teddy Bear: 13.50 inches. Tan mohair; felt pads; brown glass eyes; bellows music box activated by pressing red circle (circle replaced-original had "Music" in gold); some fur loss on forehead; circa 1950; no I. D. $1,500

Orsi Bear: 9 inches. Brown mohair with clipped tan mohair snout; brown glass eyes; open felt mouth and pads; jointed head and arms; soft stuffed; blue felt bib (can also be red); circa 1950; raised button; chest tag. $550-$650

Zotty Bear: 20 inches. Frosted mohair; inset snout; apricot chest plate; glass eyes; apricot felt open mouth and paw pads; original ribbon; circa 1955; raised button; chest tag. $800 up
Zotty Bear: 10 inches. Same description. Raised button; chest tag. $425.
Zotty Bear: 8 inches. Same description. Raised button; chest tag. $350

Minky Zotty: 11 inches. Brown and tan plush; open felt mouth; clipped mohair muzzle; glass eyes; raised script button; chest tag; original ribbon; 1950s. $400

Cosy Teddy: 7 inches. Brown and white dralon; tan muzzle; open felt mouth and paws; soft stuffed; circa 1955; raised button; bear head chest tag. $125-$150

Cosy Teddy: 16 inches. White plush with tan chest plate; open felt mouth and pads; glass eyes; circa 1958; bear head chest tag. $350

Jackie Bear: 13 inches. Cream mohair; glass eyes; felt pads; excelsior stuffed; all jointed; pink stitch across nose; airbrushed belly button; issued with a booklet to celebrate the company's jubilee. 1953; raised script button. This bear has sold for more than $10,000 with a chest tag in tissue mint condition.

Teddy Bear: 6 inches. Tan mohair; glass eyes; circa 1950; raised button. $250-$300

Teddy Bear: 8 inches. Bright gold mohair; glass eyes; original ribbon; circa 1955; raised button; chest tag. $350-$375.
Teddy Bear: 3.50 inches. Gold mohair; black glass eyes; raised button; U S Zone tag; circa 1948. $350-$375.

Teddy Bear: 6 inches. Tan mohair; glass eyes; no paw pads; original ribbon; circa 1955; raised button. $350

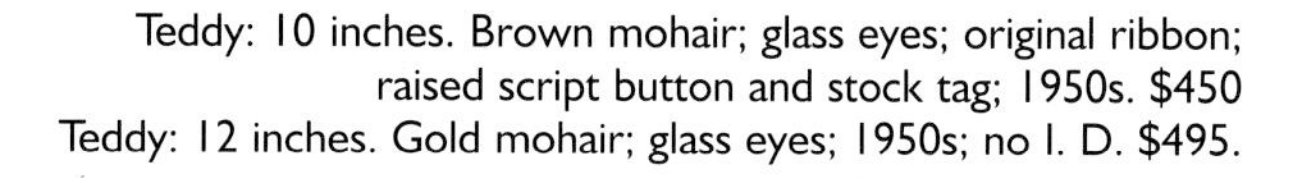

Teddy: 10 inches. Brown mohair; glass eyes; original ribbon; raised script button and stock tag; 1950s. $450
Teddy: 12 inches. Gold mohair; glass eyes; 1950s; no I. D. $495.

Teddy Bears: 8 inches and 13 inches. Brown and gold mohair; circa 1950; glass eyes. Brown: raised button. Gold: chest tag. $450-$550

Teddy Bear: 13 inches. White mohair; glass eyes; original ribbon; circa 1950 No I. D. $650

Teddy Bear: 13 inches. Gold mohair; glass eyes; squeaker; circa 1955. All I. D. U S Zone tag. $650

Teddy Bear: 14 inches. White mohair; glass eyes; original ribbon; circa 1950; label on foot "Made in Germany for B. Altman Co. N. Y." $900 up

Original Teddy Bear: 17 inches. Caramel mohair; glass eyes; circa 1955; raised button; chest tag. $600 up

Teddy Bear: 20 inches. Cinnamon mohair; glass eyes; circa 1950; raised button. $1,400

Teddy Bear: 26 inches. Tan curly mohair; glass eyes; circa 1950 No I. D. $3,100

Teddy Bear: 10 inches. Tan mohair; glass eyes; circa 1965; dressed by F. A. O. Schwarz; holding city mouse also dressed by the store. Schwarz hang tag. $450-$500

Original Teddies: 7 inches. Beige and caramel mohair; shaved snout and eye area; plastic eyes; 1960s; incised buttons; bear head chest tags. $195

Original Teddy: 14 inches. Tan mohair; glass eyes; shaved muzzle and eye area; incised button; chest tag. 1960s. This style of bear produced for more than 20 years. $475

Original Teddy: 9 inches. Blond mohair; clipped muzzle; plastic eyes; original ribbon; circa 1960; raised script button; bear head chest tag. $225

Petsy Bear: 14 inches. Blonde plush; glass eyes; velour foot pads; wool and rubber stuffed; ribbon; circa 1960 Raised script button; bear head chest tag. $350

Papa Bear: 18 inches. Boxed with certificate. Gold mohair. The first limited edition made by Steiff for collectors. 1980 $900 up *Mama and Baby*. Boxed with certificate; held together with orange ribbon. 1981. $900 up

Above: Floppy Zotty: 13 inches. Tan mohair; open mouth; embroidered sleep eyes; soft stuffed; circa 1960 No I. D. $125

Left: Zooby Bear: 11 inches. Brown mohair with darker brown clipped feet; plastic eyes; open felt mouth and claws; swivel head and bent arms; stands on vinyl soled feet; 1964. Raised button. $900 up

Toldi Bear: 15 inches. Dressed plush bear; plastic eyes. 1974. $175. *Toldi*: 11 inches; dressed plush. 1980s. $110 Both have split chest tags and brass buttons.

Margaret Strong Bear: 12 inches. Gold mohair; plastic eyes; this bear was introduced at the Strong Museum in Rochester, New York, when it opened in 1982. Hans Otto Steiff, Margarete's great grand nephew, made an appearance to sign the foot pads. $400 up

Zotty Bear: 16 inches. Brown tipped mohair; apricot chest plate; inset muzzle; plastic eyes; unjointed; original ribbon; incised button; split chest tag; 1972. $225
Zotty: 20 inches: Plush; plastic eyes; velour pads; brass button; split chest tag. 1980s. $350

Petsy Bears: 10 inches; plush; plastic eyes; velour pads; sitting bear has incised button; 1970s. $275 Standing bear has a brass button; both have split chest tags; 1980s. $175

Richard Steiff Teddy Bear: 12 inches. Gray mohair; plastic eyes; 1982 replica, issue of 10,000; M. I. B. brass button; hang tag on paw, showing Richard Steiff's photo. $450 up

Nimrod set: dressed mohair bears; replica of the Teddy Roosevelt Commemorative set first introduced in 1953. Ltd. ed. of 10,000 1983. $450 up

Klein Archie: 14.50 inches. Brown mohair made for The Enchanted Doll House in Vermont; boxed with certificate; brass button; split chest tag. 1983. $425

Teddy: 7 inches. Gold mohair; limited edition; wears green hat and scarf; 1980s; brass button; hang tag. $95 up

Manschli Bear: 8 inches. Caramel mohair; plastic eyes; soft stuffed; swivel head; brass button; split chest tag. 1983. $295

Dicky Bear: 13 inches. Reissue of 1930 bear in box with certificate; 1985. $190-$225
Papa Bear: 17 inches. The first of limited edition bears; made for U. S. and German markets; in box with certificate; 1980 $900 up
Clown Bear: 13 inches. Reissue of 1926 bear in box with certificate; 1980 $350

Petsy Bears: 10 inches and 33 inches. Plush; plastic eyes; brass buttons; split chest tags; 1980s. $125 and $600

Margaret Strong Cream bear: 11 inches; one of five sizes; cream mohair; brass button; 1984-1986. $375

Margaret Strong Chocolate set: 7, 10, 12 and 16 inches. Brown mohair; leather paw pads; brass buttons; white stock and chest tags; ltd. ed. of 2,000; boxed; 1983. $650

Giengen Bear: 15 inches. Gray mohair; one of four sizes; replica of a 1906 teddy; brass button; chest tag features a unicorn (symbol of Giengen); 1985. $495

Mr. Cinnamon Bears: 10, 13 and 18 inches. Cinnamon mohair; plastic eyes; produced in recognition of a children's book; brass buttons and story booklet; 1984. $325, $475, $600

Luv Bear: 9 inches. Original teddy dressed in yellow felt vest with a red heart; brass button; split chest tag. 1985. $175

F. A. O. Schwarz Bear: 12 inches. Brown mohair; made for the toy store in 1987; F. A. O. Schwarz tag and certificate. $395

Jackie Bears: 6 inches and 9 inches. Reissue of *Jackie*, the 1953 teddy; boxed with certificate and replica booklet. 1987-1989. $250 and $350

Berryman Bear: 13 inches. Made to commemorate the 85th Anniversary of Clifford Berryman's cartoon. 1987. $450

Teddy Clown: 12 inches. Reproduction of the 1930s example. 1986. $350

Mrs. Santa Claus: 8 inches. White mohair; dressed in red print dress, hat and shoes; created especially for Hobby Center Toys (now The Toy Store) in Toledo, Ohio. Brass button; hang tag. 1987. $350 up

Bear Dollys: 12 inches. 1987 reproduction of an earlier bear produced in three colors of mohair; two of which are shown; yarn ruffs; brass buttons; white stock tags; hang tags. $295

Petsy Bear: 13 inches. Reissue of the company's brass Petsy; gold mohair with white stock tag; chest tag is a replica of the 1927 one; 1989; M. I. B. $350

Alfonzo: 13 inches. Red mohair; satin Cossack; replica of bear once belonging to Princess Xenia of Russia (relative of the Romanovs). Produced for "The Bears of Witney" in England. Ltd. ed., 1990. $600+

Teddy Baby: 6 inches. Apricot mohair; velour muzzle; plastic eyes; collar and bell; ltd. ed. of 5,000 1990s. $250 up

Musical Bear: 10 inches. Cinnamon mohair; key wound music box plays Mozart's Lullaby. Ltd. ed. of 2,000 for Harrod's, London. 1990. $450

Teddy: 13 inches. Cinnamon mohair. Made for Hamley's Toy Store, London, in 1990; certificate. Dennis Yusa collection. $400

Musical Clown: 14 inches. Cinnamon mohair. Made for Harrod's, London, in 1992. Ltd. ed. of 2,000. $400

Ralph Lauren Bears: 13 and 17 inches. Made in 1991 and 1992 for Lauren. Dressed in replicas of adult clothing. $400 and $800

Teddy Bear: 15 inches. Ltd. ed. of a 1926 bear; tipped mohair; large eyes; white ear and chest tags; M. I. B. 6,000 in 1991; brass button. $600

Otto Bear: 15 inches. Made for U. S. only; replica of 1912 and bearing Otto Steiff's name; commemorating the establishment of the Steiff Company with America, hence the date on one foot and flag on the other. Ltd. ed. of 5,000; 1992; brass button; white stock tag. $500

Amelia Teddy: 12 inches. Pink mohair; glass eyes; wearing beautifully made leather jacket and hat, goggles and scarf; made in a limited edition of 650 for I. Magnin Stores; 1993. White tag on chest; brass button. $625 up

Musical Teddy: 15 inches. Green mohair; brocade waistcoat; key wound music box plays Rossini's "The Thieving Magpie." Ltd. ed. of 2,000 for Harrod's, London. 1993. $550

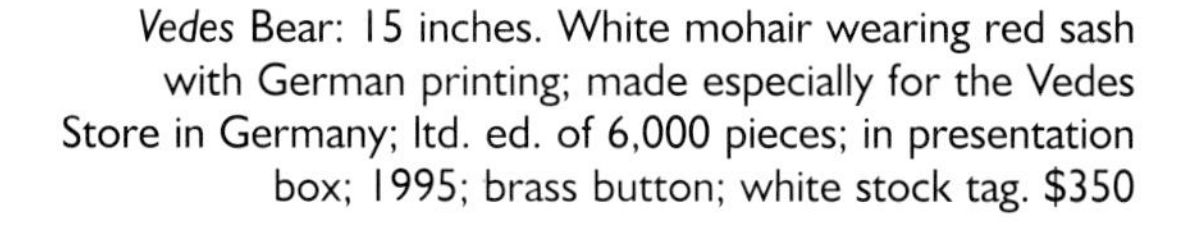

Vedes Bear: 15 inches. White mohair wearing red sash with German printing; made especially for the Vedes Store in Germany; ltd. ed. of 6,000 pieces; in presentation box; 1995; brass button; white stock tag. $350

Steel Worker Willi: 14 inches. Mohair; cotton pants; leather helmet, apron and boots; metal ladle; ltd. ed. of 1,500; 1996. Made for store in Germany. Brass button; white stock tag. $500

Passport Bear: 14 inches. North American exclusive; 1997; complete with suitcase; boxed; brass button; split chest tag. $395 at issue.

Reginald and Ghurka Back Pack: 15 inches. Safari Bear complete with suit, pith helmet; full sized Ghurka back pack; safari book relating his adventures; various pins and a cloth carrying case; 1997; buttons and tags. $950

Carrying bag for *Reginald*

CHAPTER III

TEDDY BEAR SPECIALTIES

THESE TEDDIES CAN BE THE SUNSHINE OF YOUR LIFE

The Steiff Club has its roots in Europe, but was quickly ear-marked by Steiff USA. This special club department, now run by Rita Bagala, introduces one or two yearly products available to members only. Membership includes a welcome gift and a quarterly newsletter. The periodical apprises members of special events and toys (available only to them) as well as answering questions they might have about their collections.

The Toy Store in Toledo, Ohio, has held a Steiff Festival every year since 1986. At each event a teddy or other collector toy is offered in a limited edition to commemorate the occasion. A Steiff Convention in Germany was premiered in 1997 to celebrate Margarete's 150th birthday and this gala has continued each year since and also introduces a special event product. Disney World holds a yearly Bear and Doll convention in which Steiff holds a special place. Many shops and entrepreneurs host happenings that feature a Steiff piece as the "special of the house."

Steiff Club Publication for the summer of 1999, showing the Millennium Bear for club members.

Sam Club Bear: 10 inches. Gold mohair; made in 1993 for U. S. Steiff Club members only; boxed with certificate; brass button; white stock tag; ceramic disc hanging from red, white and blue ribbon. $500

Clown Club Bear: 10 inches. Tipped mohair; wearing a clown hat and ruff; made in 1993 for U. S. Steiff Club members only; boxed with certificate; brass button; white stock tag; ceramic disc. $450

Blue Club Bear: 13 inch. Replica of Eliot, the 1908 blue bear that sold at Christie's, London, in 1993 for $74,250 Exclusive to club members in 1994. $475

Club Teddy: 3.50 inches. Blue teddy given as a premium to Steiff club members in 1996. $100

Picnic Club Bear: 14 inch. 1997; available to Steiff Club members only; complete with basket; "wine" bottle; tablecloth and two china plates; ceramic pendant; mint in box with certificate; brass button; white stock tag. $375

Mr. Vanilla Teddy Bear: 10 inches. White mohair; limited edition of 1,000 made expressly for Hobby Center Toys (now the Toy Store) of Toledo, Ohio; 1989. Signed on foot by Jörg Juninger; brass button; replica chest tag. $600 up

Dicky Bears: 6 inch. Rose and mauve mohair; plastic eyes; velour pads; collar; ltd. ed. of 2,000; made for the Toy Store in Toledo, Ohio (formerly Hobby Center Toys); 1991; brass button; white stock tag; chest tag; hang tag. $300 each

Teddy Baby Rosé: 7 inch. Pink mohair; plastic eyes; collar and bell. Limited edition of 1,000 for Hobby Center Toys (now the Toy Store) in 1990. $450 up

T. R. Bear: 7 inch. Tipped brown mohair; felt jacket and pith helmet; wire rimmed glasses; ltd. ed. of 1,500 pieces; made for the Steiff Convention held in Toledo, Ohio, in 1994 expressly for the Toy Store. Brass button; white stock tag with red printing; chest tag. $275

Event Bear: 8 inch. Made for 1996 Festival of Steiff. Ltd. ed. of 100 pieces; wears ruff, banner, and includes a ceramic clown mask to commemorate the circus theme. 1996. Brass button; split chest tag. $295

Margarete Steiff Jubilee Bears: 11 inches. Gold, brown and white mohair; made to celebrate the founders 150th birthday in 1997. Wears a porcelain disc with Margarete's photo on it. Brass button; white stock tag; disc with ribbon; paper tags. $225 each

Blackey: 13 inches. Black bear with blue eyes made in a limited edition to celebrate the first Steiff Convention held in Giengen, Germany. 1997. $650 up
Additional convention bears include:
Whitey in 1998. $595
Rosey in 1999. $595

U. F. D. C. Luncheon Bear: 9 inches. Made by Steiff to celebrate the United Federation of Doll Club's 40th birthday in St. Louis, Missouri. Ltd. ed. of 260 in 1989; wears red ribbon with data; brass button; stock tag; hang tag on foot. Donna Felger collection. $350 up

Teddy: 9 inch. Tan mohair; plastic eyes; wearing Roman toga and laurel wreath; made for European market wherever there was a Roman occupation; each area has a different Emperor. This is "Imperator Vespasian" of Giengen, Germany; 1994; brass button; hang tags. $325

Tinsel Teddy Bear: 4 inch. White mohair; all jointed; wearing red knit sweater and hat; limited edition of 2,500 pieces; the 1st issue in a collection of historic miniatures conceived by Steiff U. S. A.; tree ornament for 1994; comes in box with scrolled certificate; brass button; hang tag; red label on sweater. $250 up

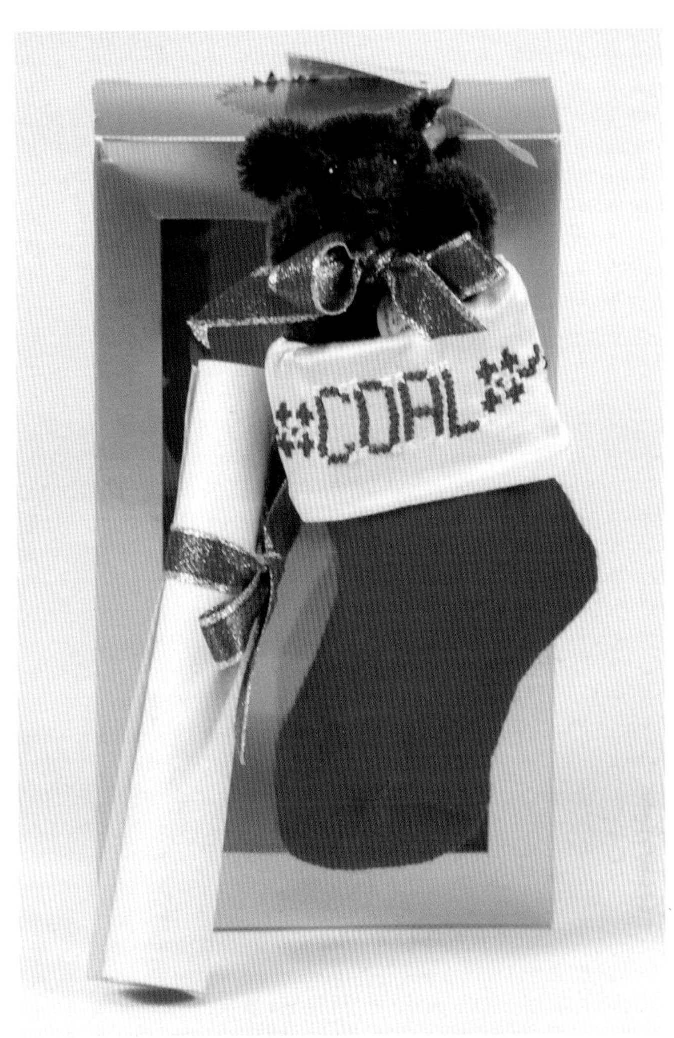

Coal: 3.50 inches. Mohair bear; in stocking with embroidery; certificate; made in a limited edition for Christmas 1995; brass button; split chest tag. $225

Teddy: 12 inch. 1990 Disney World Convention bear. Signed on foot by Jörg Juninger; limited edition of 1,000 Ribbon; enamel pin; brass button; white stock tag. $575 up

Snowman Ornament: 4.50 inches. Mohair snowman ornament for Christmas 1998. $95

CHAPTER IV

THESE BEARS ARE MADE FOR WALKING

Bears have the ability to stand upright and apparently are easily trained to do so. Years ago, in Europe, they were common as street performers and, of course, many plied this ability in circus animal acts. A bruin also rises to the occasion when feeling protective or threatened. The natural position is, however, to walk on all four feet. Grisly, polar and pandas in this realistic pose have been part of Steiff's program since the beginning.

Mechanical Bear: 9 inches. Brown mohair; tan inset muzzle; glass eyes; the head moves by levering the tail; printed FF button; circa 1930. $2,800 up

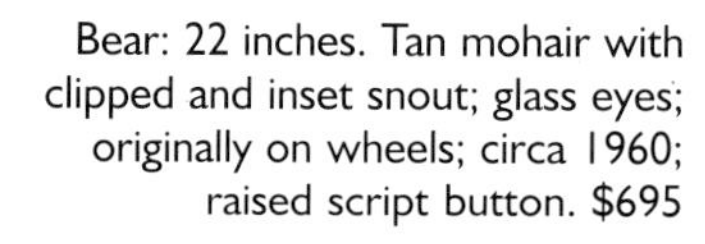

Bear: 22 inches. Tan mohair with clipped and inset snout; glass eyes; originally on wheels; circa 1960; raised script button. $695

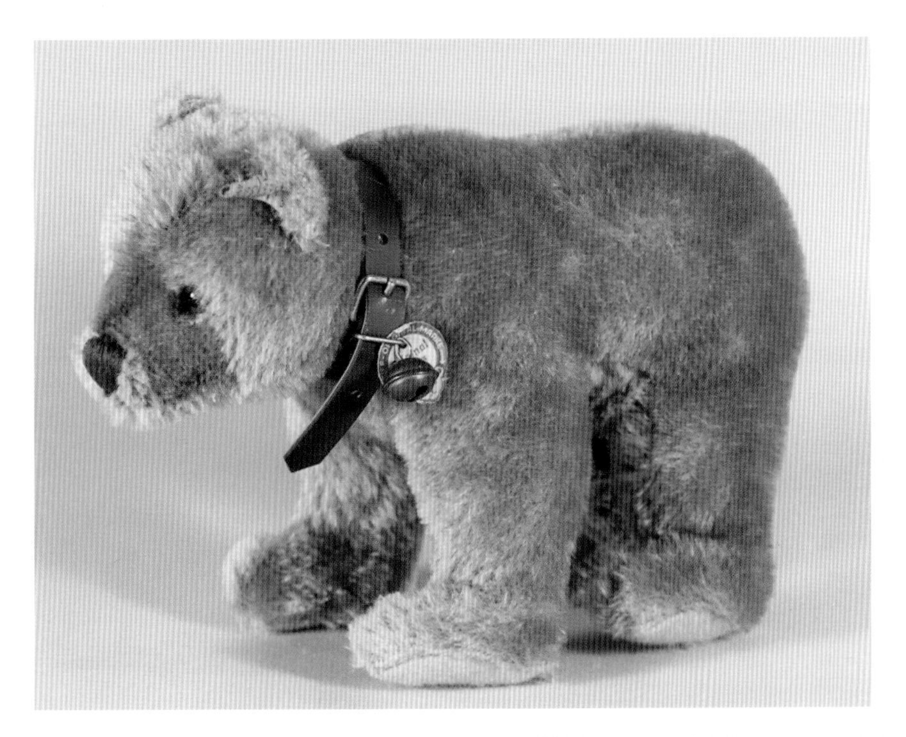

Young Bear: 7 inches. Caramel mohair; glass eyes; 1950s; raised button; chest tag. $250 up

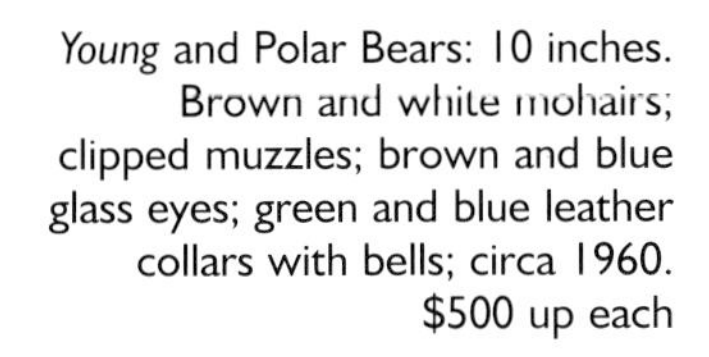

Young and Polar Bears: 10 inches. Brown and white mohairs; clipped muzzles; brown and blue glass eyes; green and blue leather collars with bells; circa 1960. $500 up each

Polar Bear: 6 inches. White mohair; glass eyes; original collar and bell; circa 1965; incised button; F. A. O. Schwarz tag. $225

Above: Replica Polar Bear: 8 inches. White mohair; jointed legs; ball and socket jointed neck; limited to 3,000 pieces; 1987; one of museum series. Boxed with certificate; brass button. $400 up

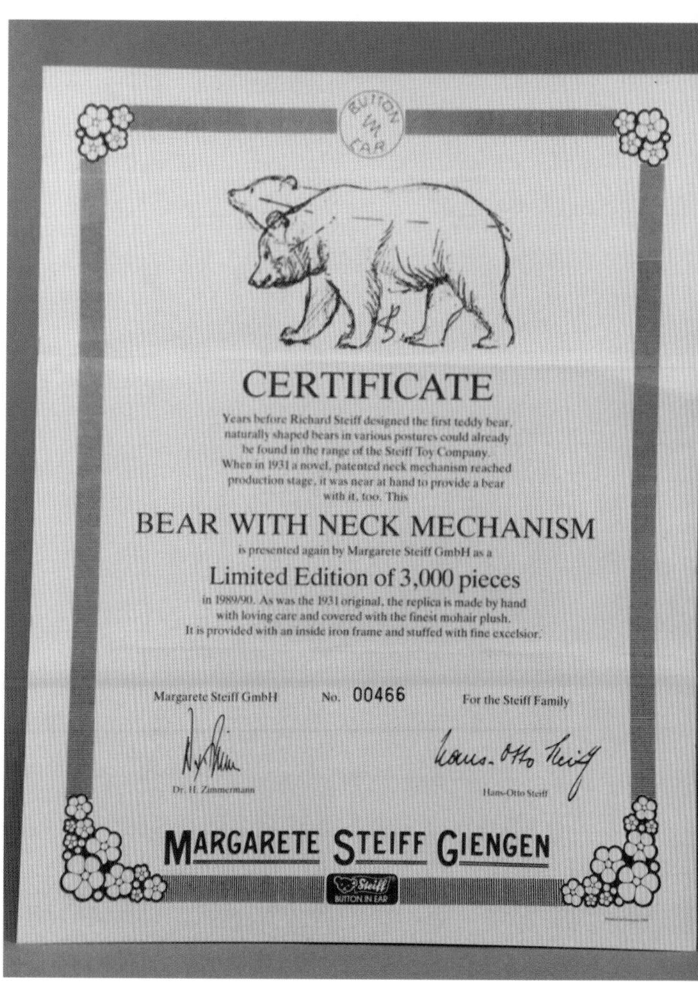

CERTIFICATE

Years before Richard Steiff designed the first teddy bear, naturally shaped bears in various postures could already be found in the range of the Steiff Toy Company. When in 1931 a novel, patented neck mechanism reached production stage, it was near at hand to provide a bear with it, too. This

BEAR WITH NECK MECHANISM

is presented again by Margarete Steiff GmbH as a

Limited Edition of 3,000 pieces

in 1989/90. As was the 1931 original, the replica is made by hand with loving care and covered with the finest mohair plush. It is provided with an inside iron frame and stuffed with fine excelsior.

Margarete Steiff GmbH — No. 00466 — For the Steiff Family

Dr. H. Zimmermann — Hans-Otto Steiff

MARGARETE STEIFF GIENGEN

Steiff BUTTON IN EAR

Right: Certificate included with Polar Bear, showing and explaining neck mechanism.

Tapsy Bear: 8 inches. Brown and white plush; black and white plastic eyes; soft stuffed; velour pads; this same bear is presented on wheels and called *Rattler*; 1970s; incised button; split chest tag. $150

Panda Bear: 6 inches. Black and white mohair; glass eyes; swivel head; original collar and bell; circa 1958; raised button. $225

CHAPTER V
CATS
PUSSYCAT, PUSSYCAT WE LOVE YOU.

Steiff's first catalog was published in 1892 and cats were an integral part of their program. A striped tabby on wheels holding a mouse in its mouth and a sitting tabby were illustrated and available in several sizes. Data from that same reference material shows that cat skittles, cats riding on velocipedes and as pin cushions were also offered.

Real cats have been worshipped, endowed with mystical powers and above all proven to be the pet of choice for millions of devotees the world over. It is no wonder, then, that each and every year the company has produced felines in every form to satisfy the demand.

Cat: 6 inches. Tan felt airbrushed with charcoal stripes; black glass eyes; has yarn ball between feet; neck cord with bell; circa 1910; no I. D. $1,500

Velvet cat pin cushion; 4 inches. Faded striped velvet; shoe button eyes; circa 1910; printed FF button. $500 up

Six way jointed Cat: 9 inches. White mohair; tri-colored glass eyes; pink single strand nose, mouth and claws; felt ears; all jointed including tail; circa 1908; blank button. $900 up

Six way jointed Cat: 9 inches. White and gray striped mohair; glass eyes; pink floss nose, mouth and claws; jointed at head, legs and tail; squeaker; circa 1910; printed FF button; full white stock tag. $900 up

Cat: 12 inches. Silky gray mohair with faint tabby markings; green glass eyes; printed FF button; circa 1920. $325

Cat: 10 inches. Off-white mohair; glass eyes; all jointed; circa 1910; printed FF button. $900 up

Kitten: 8 inches. Striped mohair, worn; glass eyes; head moves by levering the tail; no I. D. Circa 1928. $250

Fluffy Cat: 4 inches. Faded and worn; originally off-white with blue stripes; unusual and distinctive blue/green eyes; swivel head; one of Steiff's most appealing felines; circa 1930; no I. D. $200

Fluffy Cat: 5 inches. White and airbrushed blue mohair; distinctive green mottled eyes; one of the most popular cats. *Fluffy* was produced in the late 1920s and 1930s; since this animal has a U S Zone tag, it is a possible pre-war hold over. Printed FF button; full U S Zone tag. $600 up

Tabby Cat: 4 inches. Mohair; glass eyes; original bell; chest tag; circa 1930. $500 up

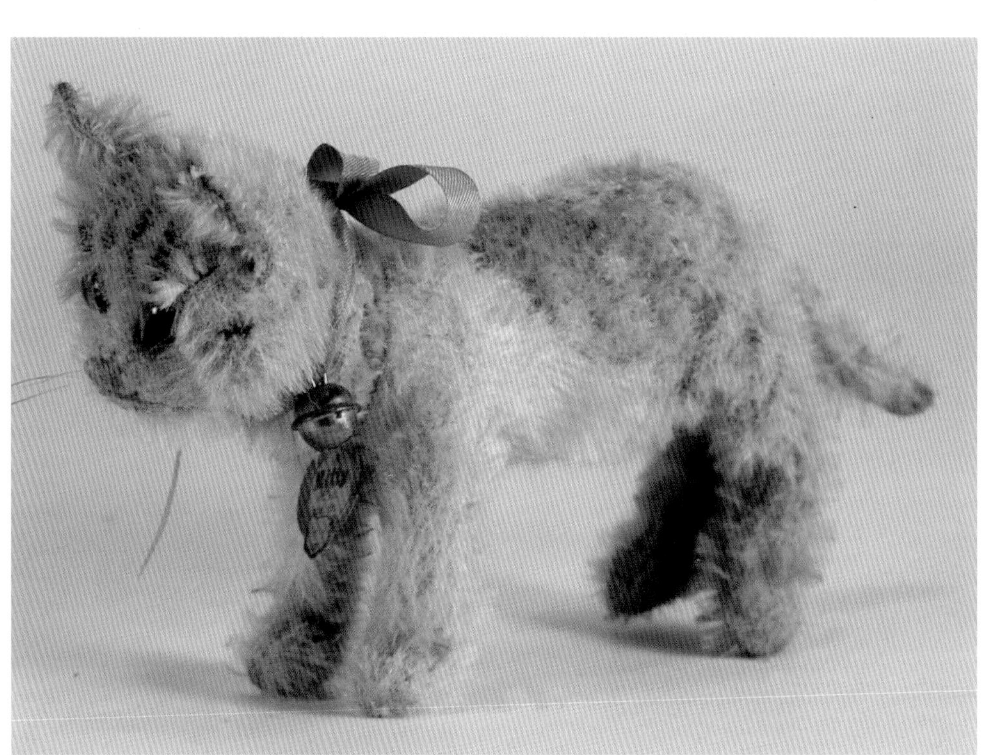

Kitty: 6 inches. Striped mohair; glass eyes; head moves via tail; shows some wear; circa 1930; square bear head chest tag. $650

Moveable head Siamese Cat: 5 inches. Cream and brown mohair; blue glass eyes; head moves in a circular fashion via tail; circa 1930; printed FF button. $850 up

Siamese Cat: 9 inches. Brown and beige mohair; blue glass eyes; excelsior stuffed; moving tail causes neck to turn; note that tail is kinked (a feature that some of this breed possesses); circa 1930; no I. D. $1,900

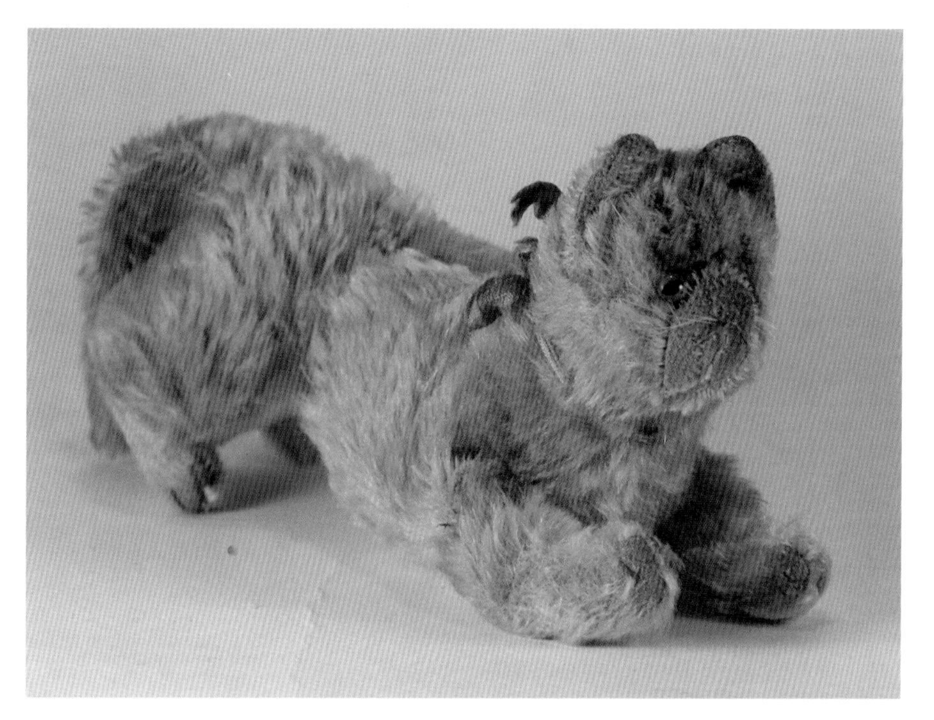

Cat: 7 inches. Orange airbrushed striped mohair; glass eyes; circa 1930; printed FF button. $700

Cat: 9 inches. Purple and white mohair (rare color); green mottled eyes (like Fluffy's); tail turns head; circa 1930; printed FF button. $2,500

Wooly Cat: 2 inches. Calico wool ball cat; swivel head; glass eyes; curled tail; original ribbon; circa 1935; printed FF button. $300

Susi Cat: 6 inches, Off-white striped mohair; green glass eyes; swivel head; 1940s; square bear head chest tag. $400

Susi Cat: 8 inches. Mohair; glass eyes; swivel head; 1940s; square bear head chest tag with brown printing. $525

Kitty: 8 inches. Striped mohair; glass eyes; all jointed; 1940s; square head bear chest tag with brown printing. $525

Siamese Cat: 4 inches. Brown and tan airbrushed mohair; open felt mouth; bright blue glass eyes; circa 1949; raised button; chest tag with *Siamy* in brown. $600-$650

Siamese Cat *Simh*: 9 inches. Airbrushed plush;. blue glass eyes; original ribbon; rare; circa 1950; raised button; chest tag. $600 up

Kitty: 8 inches. Black and white striped mohair; green glass eyes; all jointed; circa 1950; raised script button. $175

Kitty: 5 inches. Mohair; glass eyes; all jointed; 1950s; button; chest tag. $150

Gussy Cat: 6 inches. Black and white mohair; open velvet mouth; glass eyes; swivel head; circa 1955; raised script button. $195

Fiffy Cat: 13 inches. White and gray striped mohair; green glass eyes; swivel head; ribbon not original; circa 1958; raised script button. $285
Tapsy Cat: 7 inches. Cream and brown striped mohair; green glass eyes; excelsior stuffed; squeaker; bow not original; circa 1955; raised script button. $195

Musical Cat: 8.50 inches. Mohair head and arms; green glass eyes; tubular body; press down to activate music; original dress; 1950; U S Zone tag. $600 up

Tabby Cat: 3 inches. Striped mohair; glass eyes; original ribbon and bell; circa 1955; bear head chest tag. $150

Cats: 3.50 inches. Black and off-white striped mohair; green glass eyes; swivel head; ribbons; good condition; 1950s; no I. D. on sitting cat; raised button on standing cat. $110 each

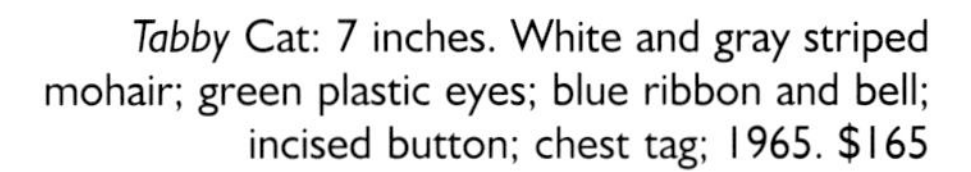

Tabby Cat: 7 inches. White and gray striped mohair; green plastic eyes; blue ribbon and bell; incised button; chest tag; 1965. $165

Kitty Cat: 4.50 inches. Striped mohair (earlier version that has stripes totally across the back; later example has space not striped in the middle of the back); glass eyes; all jointed; original ribbon; 1950s; raised button; bear head chest tag. $200

Cosy Kitty; 7 inches. Gray and white plush; green glass eyes; soft stuffed; circa 1959; raised button; bear head chest tag. $125

Snurry Cat: 8 inches. Mohair; sleep eyes; circa 1955; chest tag. $395

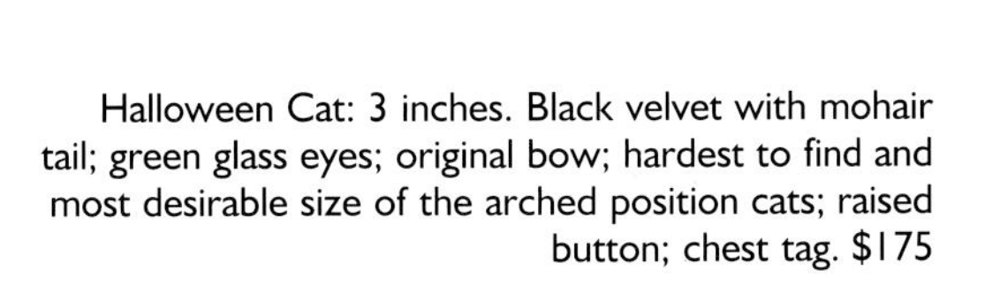

Halloween Cat: 3 inches. Black velvet with mohair tail; green glass eyes; original bow; hardest to find and most desirable size of the arched position cats; raised button; chest tag. $175

Wool ball Cat: 2.75 inches. Woolie with jointed head; glass eyes; no I. D. $30
Wool ball Cat: 2 inches. Woolie with jointed head; glass eyes; original ribbon. $40

Cosy Siam Cat: 9 inches. Realistically shaded dralon; bright blue glass eyes; soft stuffed. Siamese cats are harder to find than others. Late 1970s; brass button; split chest tag; hang tag. $175

Lizzy Cat: 6 inches. Orange and white plush; green plastic eyes; note straight up tail; circa 1965; incised button; chest tag. $125

Susi Cat: 5 inches. Sitting gray and white striped tabby; green eyes; swivel head; original ribbon; incised button; 1960s. $125

Tabby cat: 6 inches. Gray and white airbrushed mohair; green plastic eyes; pink floss nose and mouth; claws are airbrushed; excelsior stuffed; original ribbon; 1978; incised button; chest tag. $115-$120
Susi Cat: 7.25 inches. Same description except sitting position; glass eyes; incised button; chest tag. $130-$135
Tabby Cat: 7 inches. Same description as smaller; incised button; chest tag. $140

Lizzy Cat: 9 inches. White and gray striped mohair; upright tail; glass eyes; original ribbon; 1972; incised button; split chest tag. $175
Lizzy Cat: 6 inches. Same description as above; circa 1968; incised button; bear chest tag. $150

Cattie Cat: 8 inches. Apricot striped plush; plastic eyes; soft stuffed; original bow; circa 1974; incised button; split chest tag. $90

CHAPTER VI

DOGS

ELVIS SANG ABOUT A HOUND DOG AND MADE IT FAMOUS, BUT STEIFF MADE MANY MORE DOGS OTHER THAN BASSETS.

I can't, off-hand, think of a single dog that Steiff has not fabricated. And made well, I might add. Their dogs are extremely popular not only among collectors, but to breeders and owners of the live versions. Such a person tends to be impressed by this firm's attention to the most minute detail and usually ends up owning their doggie in several styles.

There are some specimens such as *Molly* and *Bully* who have appeared, in one form or another, periodically. Since the Bulldog is Yale's mascot, one model even came equipped with a logo blanket. Whatever your desire, you are sure to find it among Steiff's kennel club.

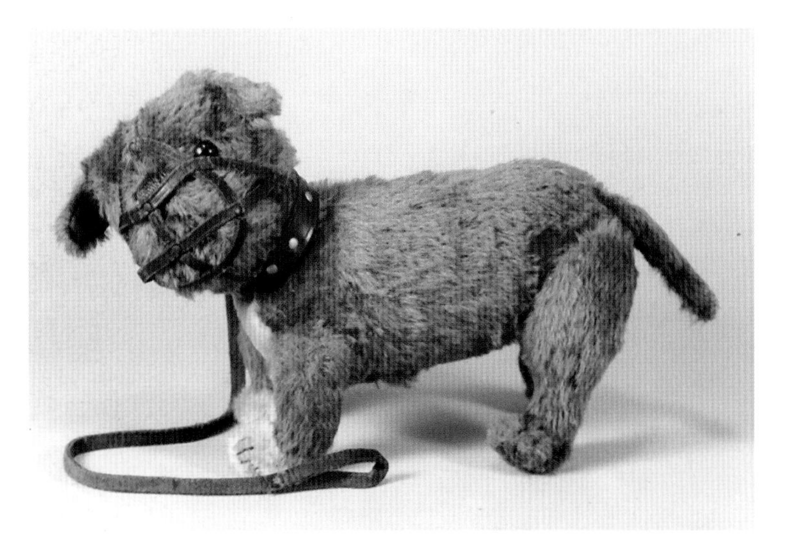

Tige Bulldog: 9 inch. Rust mohair with cream chest and front feet; shoe button eyes; all jointed; working squeaker; original leather muzzle and leash; circa 1911; No I. D. $900 up

Dog: 6 inches. Felt with airbrushed spots; shoe button eyes; moth holes professionally mended; circa 1912; no I. D. $125

Pinscher *Caesar*: 10 inches. Long off-white mohair; brown mohair ears lined with black fabric; glass eyes (airbrushed around them); all jointed; original collar. *Caesar* was the favorite pet of England's King Edward VII. This is reflected in the unusual collar tag. Rare; 1910; round brass tag with "Steiff Caesar" in raised letters; the reverse has "E. VII/1910." $2,000 up

Caesar's brass tag.

Caesar's tag on the reverse side.

Spitz Dog: 9 inches. Cream mohair; felt face, ears and legs; shoe button eyes; no I. D.; circa 1910. $250 up

Bonzo: 12 inches. Gold velvet with airbrushed spots, feet, nose and mouth; one black ear; all jointed; side pushed music box; only 115 examples of four designs were made; some had googly eyes; only the musical ones had set-in blue glass eyes; printed FF button; 1927. $4,000 up

Dog: 12 inches. Off-white mohair; one orange ear; shoe button eyes; 1920s; printed FF button. $1,250

Velvet Dog: 4 inches. Cream with brown airbrushed markings; glass eyes; original collar; printed FF button with trace of orange stock tag; 1920s. $600 up

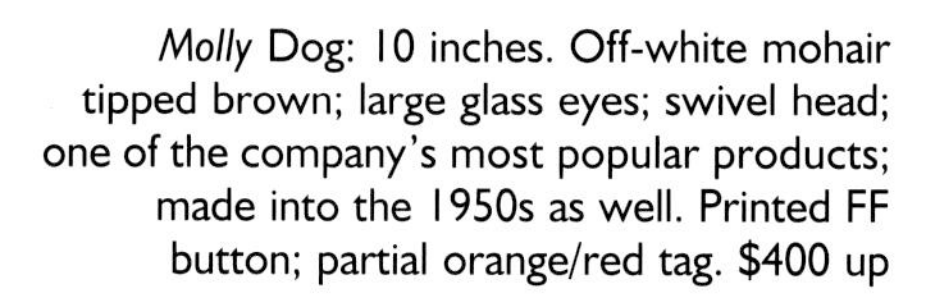

Molly Dog: 10 inches. Off-white mohair tipped brown; large glass eyes; swivel head; one of the company's most popular products; made into the 1950s as well. Printed FF button; partial orange/red tag. $400 up

Molly Dog: 5 inches. Off-white mohair; glass eyes; swivel head; circa 1925; printed FF button; hang tag is white metal rimmed and reads *Molly.* $450 up

Above: Police Dog: 7.50 inches. Mohair; glass eyes; replaced collar; circa 1925; No I. D. $350

Left: *Strupp* Terrier: 13 inches. White and black mohair; glass eyes; swivel head; circa 1925; FF button; trace of orange stock tag. $950

West Highland Terrier: 5 inches. Off-white mohair; glass eyes; swivel head; original collar and bell; circa 1920; printed FF button on collar. $375

Bully Dog: 10 inches. White and orange mohair; velvet muzzle; large glass eyes; swivel head; missing ruff or collar; circa 1930; printed FF button; trace of orange stock tag. $800 up

Charlie Purse: 10 inches. Gold mohair (shows wear); glass eyes; swivel head; zipper back; carry strap; circa 1930; no I. D. $300

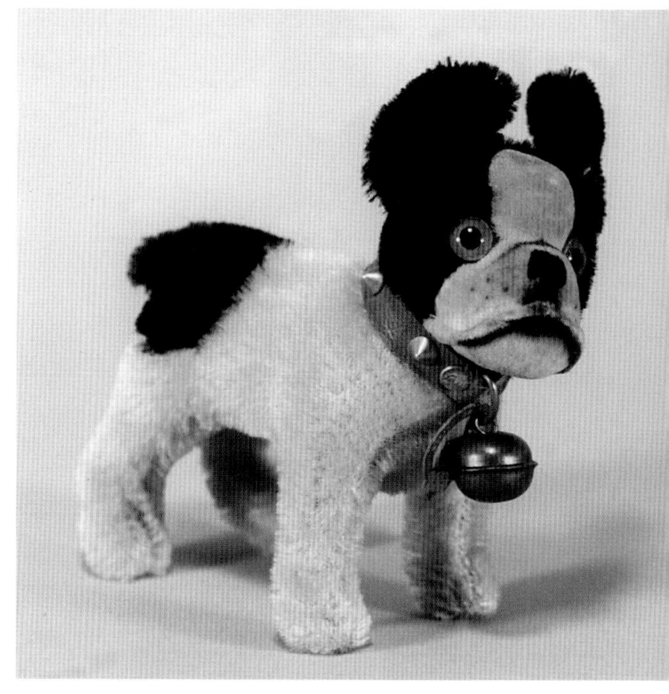

Bully Dog: 4 inches. Black and white mohair; glass eyes; swivel head; original collar with two buttons; printed FF button; trace of orange stock tag; chest tag; 1930s. K. Eschborn collection. $550 up

Chin Chin Dog: 4.50 inches. Gold mohair; velvet muzzle and feet; glass eyes; swivel head; original silk ribbon; circa 1930; printed FF button; full orange/red stock tag; square head bear chest tag. $1,100 up

Cocker Spaniel: 9 inches. Cinnamon and white mohair; glass eyes; swivel head; circa 1930; no I. D. $550

Dog *Pip*: 6 inches. Velvet faded from blue to green; tri-colored glass googly eyes; felt tongue; horsehair ruff; circa 1935; FF button; full orange stock tag. $895

Scotty Dog: 9.50 inches. Gray grizzled mohair; brown glass eyes; black floss nose; pink floss claws; swivel head; original leather collar and bell; circa 1930; printed FF button plus two buttons on collar. $800

Scotty Dog: 10 inches. Gray mohair that has faded to pink; glass eyes; swivel head; excelsior stuffed; squeaker; black and white leather collar; circa 1930; no I. D. $800

Spitz Dog: 9 inches. Off-white mohair; glass eyes; original collar; circa 1930; no I. D. $600

Scotty Dog: 9 inches. Black mohair; felt lined ears; brown glass eyes; swivel head; original collar; side pushed squeaker; circa 1930; printed FF button; square head bear chest tag. $800 up

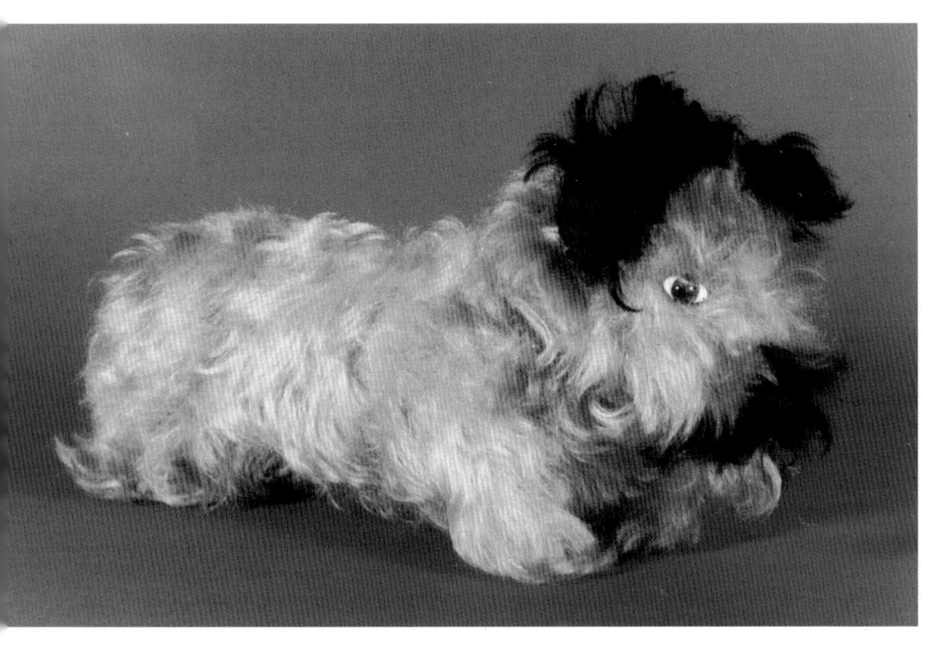

Skye Terrier: 13 inches. Long curly tan mohair; black ears and at end of muzzle; humanized eyes with whites at corners; original collar; circa 1930; printed FF button. $1,200 up

Treff: 4 inches. Tan mohair; glass eyes; swivel head; circa 1930; printed FF button; square head bear chest tag. $700 up

Sealyham: 8 inches. White mohair; glass eyes; leather collar; circa 1930; printed FF button. $750 up

Treff Purse: 11 inches. Showing the zippered back. Printed FF button; trace of orange stock tag; 1930s. $1,200

Treff Dog: 7 inches. Tan mohair; glass eyes; swivel head; circa 1930; some wear; no I. D. $300

Treff: 14 inches. Tan mohair; glass eyes set in head so that he appears to have eyelashes; swivel head; circa 1930; FF button; trace of orange stock tag. $1,000 up

Fellow Airdale: 13 inches. Silk plush; glass eyes; collar; circa 1940; FF button; square head bear chest tag with brown lettering. $500+

Airdale *Fellow*: 10 inches. Airbrushed tan and black mohair; glass eyes; squeaker; red leather collar; 1949; no I. D. $300

Beppo Dog:5 inches. Tan and black mohair; glass eyes; open felt mouth and tongue; one ear raised; all jointed; original collar; circa 1948; raised button; bear head chest tag with brown lettering. $275

Fox Dog: 9.50 inches. Wired haired terrier; somewhat later examples are referred to as "Foxy;" white, black and tan mohair; glass eyes; original collar; circa 1946; raised button; square bear head (a.k.a. Watermelon bear head) chest tag. $350

Cockie: 13 inches. Long and short gold mohair; open velvet mouth; glass eyes; swivel head; original ribbon; circa 1948; raised script button; chest tag with brown printing; U S Zone tag. $450

Fox Terrier: 11 inches. Airbrushed mohair; glass eyes; 1948; raised button; square head bear chest tag with brown printing that reads "Fox" (normally Foxy); U S Zone tag. $495

"Begging" Poodle: 10.50 inches. Black long and short mohair; tri-colored glass eyes; open felt mouth with tongue; swivel head and forelegs; collar; circa 1949; raised button. $700 up

Tosi Poodle: 5 inches. White wooly plush; glass eyes and nose; red leather collar; circa 1948; raised script button; bear head chest tag with brown printing. $395

Maidy Poodle: 9 inches. Black curly plush; tri-colored glass eyes; original red leather collar; circa 1949; chest tag on collar. $600 up
Basset Hound: 6 inches. Tan airbrushed plush; swivel head; green leather collar; chest tag on collar. $250

Bully Dog: 8 inches. Mohair; open mouth with teeth; glass eyes; horsehair ruff; swivel head; raised script button; circa 1948. $700 up

Scotty Dog: 15 inches. Black mohair; tri-colored humanistic eyes; swivel head; leather collar and bell; circa 1940; printed FF button. $600 up

Airdale: 4 inches. Naturally colored mohair; felt ears; glass eyes; circa 1950; raised script button. $125

Bully: 7 inches. Mohair; plastic googly eyes; felt blanket; Yale mascot; no I. D.; 1950s. $495 up

Bully Dog: 6.50 inches. Brown, white and black airbrushed mohair; velvet muzzle; glass eyes; swivel head; original collar; circa 1955; raised button. $100

Boxer: 5 inches. Mohair; glass eyes; velvet chin; original leather collar; all I. D.; 1950s. $150

Butch Cocker: 10 inches. Black and white mohair; plastic googly eyes; swivel head; red leather collar; circa 1950; no I. D. $375

Chow Dog: 5 inches. Wooly plush; velvet muzzle; glass eyes; 1950s; raised script button. $200

Sarras Boxer: 5 inches. Shaded tan mohair; velvet under mouth; glass eyes; original collar; circa 1955; bear head chest tag. $150 up

Musicanto Cockie: 7 inches. Mohair; glass eyes; open velvet mouth; swivel head; original ribbon; wind the tail to activate music "Brahm's Lullaby;" 1950s; all I. D. $650 up

Lying Cocker Spaniel: 13 inches. Black and white mohair; clipped muzzle; open velvet mouth; glass eyes; circa 1955; raised button. $200

Mechanical Cocker Spaniel: 12 inches. Lying *Cockie*: Black and white mohair; glass eyes; velvet open mouth; electrically wired by store for display purposes; wags tail when plugged in; circa 1950. No I. D. $425

Chow Dog: 4.50 inches. (Actually a Pomeranian.) White wooly plush; velvet ears and muzzle; glass eyes; original ribbon; circa 1958; raised button; chest tag. $225

Collie: 15 inches. Long and short shaded mohair; open mouth with felt tongue; plastic eyes; hard and soft stuffed; 1950s; incised button; chest tag. $250

Sitting Collie: 5 inches. Mohair; glass eyes; felt ears; circa 1950; raised button; chest tag. $175

Waldi Dachshound: 10 inches. Mohair; glass eyes; green leather collar; chest tag; 1950s. $175
Waldili: 9 inches. Mohair; glass eyes; non-removeable clothes; wooden rifle; all I. D. 1950s. $450

Hexie Dachshound: 5 inches. Mohair; glass googly eyes; collar; circa 1955; raised button; chest tag. $145
Bazi Dachshound: 8 inches. Airbrushed mohair; glass eyes; swivel head; collar; circa 1955; raised button; chest tag. $195

Foxie Dog: 6 inches. Natural shaded mohair; glass eyes; hard to find in sitting position; 1955; raised button. $450

Molly Dog: 4 inches. Off-white with rusty brown airbrushing; swivel head; glass eyes; circa 1950; raised button. $150

Molly Dog: 8 inches. White and cinnamon airbrushed mohair; glass eyes; swivel head; circa 1955; raised script button. $295

Mopsy Dog: 5 inches. Tan mohair; glass googly eyes; felt tongue; swivel head; ribbon not original; circa 1958; raised script button. $110

Snobby Poodle: 6 inches. Gray long and clipped mohair; tri-colored glass eyes; felt tongue; rope tail with pom pom; all jointed; original collar; circa 1955; raised button; chest tag. $95

Snobby Poodle: 15 inches. Long and short black mohair; red felt tongue; humanized glass eyes; lying position harder to find than the all jointed version; 1950s; raised button. $225

Snobby Poodle: 6 inches. Black wool plush clipped to "show" standard; glass eyes; original collar with chest tag *Snobby* in brown printing; raised script button; U S Zone tag; 1952. $495

Snobby Poodle: 7 inches. White wooly plush; glass eyes and nose; red leather collar; 1952; raised script button; chest tag with brown printing. $495

Pulac Poodle: 36 inches. Gray alpaca; leather nose; long dangling legs; 1950s. $1,000

Walther Poodle: 6 inches. Tan velvet with gray plush head, front body, ankle ruffs and tail plume; leather-like face with painted eyes; original collar (leather) with metal tag "Walther;"made as a promotion for the Walther Shoe Co; raised button. $600

Tessie Schnauzer: 12 inches. Gray alpaca; open velvet mouth; glass eyes; felt lined ears; swivel head; collar; 1950s; chest tag. $425 up

Bernie St. Bernard: 8 inches. White airbrushed mohair; plastic eyes; collar and wooden keg; circa 1955; raised script button. $425

Arco German Shephard: 22 inches. Airbrushed mohair; open felt mouth with tongue; glass eyes; collar; circa 1955; raised button. $575

Arco German Shephard: 14 inches. Airbrushed mohair; open felt mouth with tongue; glass eyes; collar; circa 1955; raised button; chest tag. $795

Arco German Shephard: 15 inches (plus tail). Airbrushed mohair; open felt mouth with felt tongue; plastic eyes; red leather collar; circa 1965; incised button; chest tag. $395

Corso Afghan: 10 inches (plus tail). Long mohair; airbrushed muzzle and ears; plastic eyes; raised script button; 1960s. $250

Cosy Cockie: 10 inches. Gold and white wooly dralon; glass eyes; ribbon bow; circa 1968; incised button; chest tag and hang tag. $160 *Cosy Schnauz*: 9 inches. Tan and taupe wooly plush; plastic eyes; red collar; circa 1968; incised button; bear head chest tag. $185 *Biggie* Beagle: 8 inches. Shaded mohair; plastic eyes; swivel head; red collar; circa 1968; incised button; bear head chest tag. $190

Musical *Cockie*: 12 inches. Brown and white mohair; plastic eyes; open felt mouth; wind tail to activate music box; red leather collar; circa 1960. $625

Dally Dalmation: 4 inches. Black and white mohair; plastic eyes; swivel head; vinyl collar; circa 1960; chest tag. $175

Above: *Electrola Fox*: 7 inches. White and tan shaded, sheared plush; mohair ears (one upright); brown glass eyes; collar; 1960s; raised button; chest tag. $600 up

Left: Cocker: 5 inches, Black and white mohair; plastic eyes; attached by a chain to a wooden dog house; includes bone and dish; assembled and sold by F. A. O. Schwarz in the 1960s; raised button; chest tag. $1,200 up

Cosy Molly Dog: 8 inches. Dralon plush; plastic eyes; open felt mouth; soft stuffed; original ribbon; circa 1960; raised button; chest tag. $125.

Mopsy Dog: 6 inches. Airbrushed dralon; plastic eyes; felt tongue; swivel head; soft stuffed; circa 1961; raised script button; bear head chest tag. $325 up

Peky Dog: 3.50 inches. Gold mohair; cream and black muzzle; plastic eyes; swivel head; blue ribbon; circa 1960; raised script button; bear head chest tag. $125

Pekinese: 10 inches. Gold mohair; white and black muzzle; glass eyes; swivel head; circa 1960; incised button; chest tag. $170-$175.
Snobby Poodle: 4 inches. Gray plush with mohair accents; plastic eyes; rope tail; jointed legs; circa 1960; incised button; chest tag. $85-$90

Raudi Schnauzer: 10 inches. Silvery tipped mohair; glass googly eyes; swivel head; 1960s; incised button; chest tag hung from original collar. $250

St. Bernard: 9 inches. Airbrushed mohair; glass eyes; original collar; circa 1965; incised button. $195 up

Biggie Beagle: 4 inches. Airbrushed mohair; swivel head; plastic eyes; original collar; incised button; split chest tag; circa 1973. $125

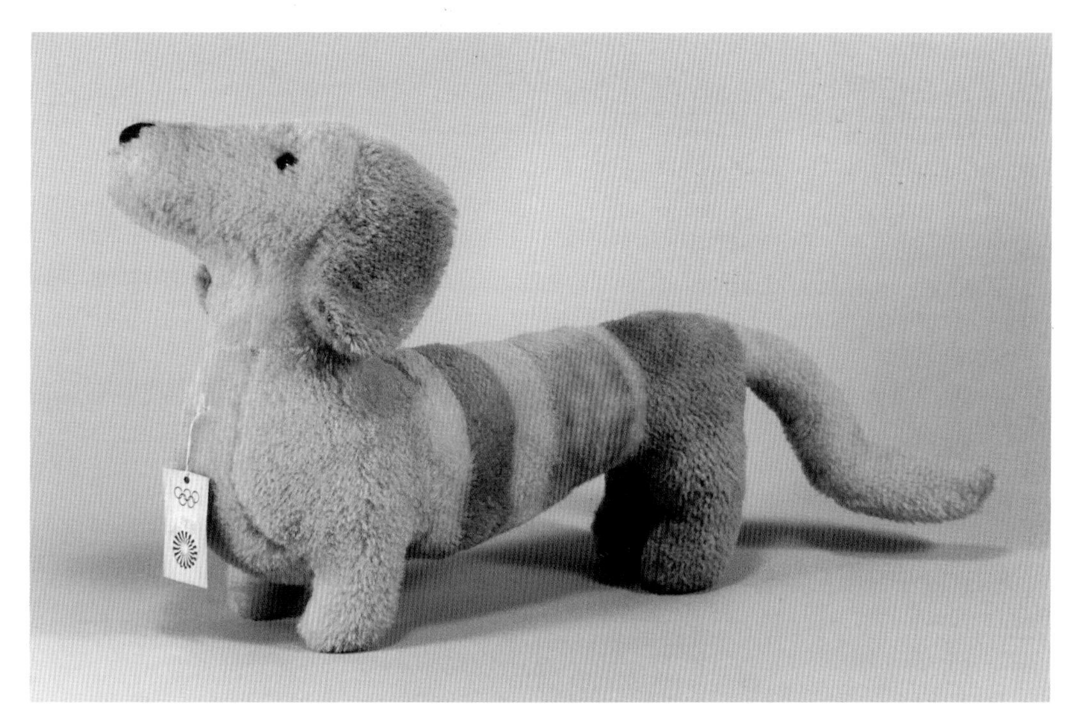

Olympic Dachshound: 13 inches. Made for the 1972 Munich Olympics; multicolored plush with ringed body stripes to represent olympic symbol; incised button; olympic tag with 5 rings. $325

Dachshound: 10 inches. Cinnamon plush; plastic eyes; excelsior and wool stuffed; vinyl nose and colar; circa 1972; incised button; split chest tag. $150

Klaff Hound: 4.50 inches. Plush; plastic eyes and nose; raised button; split chest tag; circa 1970s. $150 up

Schnauzer: 18 inches. Plush; glass eyes; leather collar; circa 1972; raised button; split chest tag. $450

Revue Susi: 10 inches. Gold mohair; black and white plastic eyes; circa 1972; (missing collar); incised button; split chest tag. $150

Puppy: 25 inches. Golden rust plush; plastic eyes; vinyl nose; soft stuffed; circa 1980; brass button. $125 up

Wheaton Terrier: 13 inches. Dralon; felt tongue; circa 1975; incised button; split chest tag; F. A. O. Schwarz tag. $450

Wolfi Dog: 16 inches. Plush; plastic eyes; soft stuffed; 1980s; brass button; split chest tag; hang tag. $345-$375

Molly Husky: 24 inches (plus tail). Grey and white plush; plastic eyes; soft stuffed; red collar; 1980s; brass button; split chest tag. $350 up

CHAPTER VII

DOWN ON THE FARM

DID OLD MACDONALD HAVE A FARM? COULD STEIFF HAVE PROVIDED THE LIVESTOCK? THE ANSWER TO THESE QUESTIONS IS READILY APPARENT.

Neighs, brays, bleats, moos. They are all music to the ears of those interested in farm animals. From the 1920s pig, whose snout was gauze and possessed a mighty side pushed squeaker, to the latest in soft plush, the barn collection has always appealed. The progression of fabrics is interesting, as well. Felt, velvet, coat wool, mohair, dacron, trevira velvet and soft man-made plush have all had their place over the years.

Horse: 12 inches. Brown coat wool; shoe button eyes; felt ears; horsehair tail; leather harness; all jointed; circa 1910; no I. D. $2,000

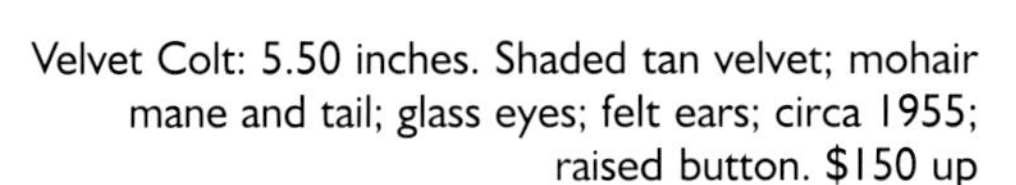

Velvet Colt: 5.50 inches. Shaded tan velvet; mohair mane and tail; glass eyes; felt ears; circa 1955; raised button. $150 up

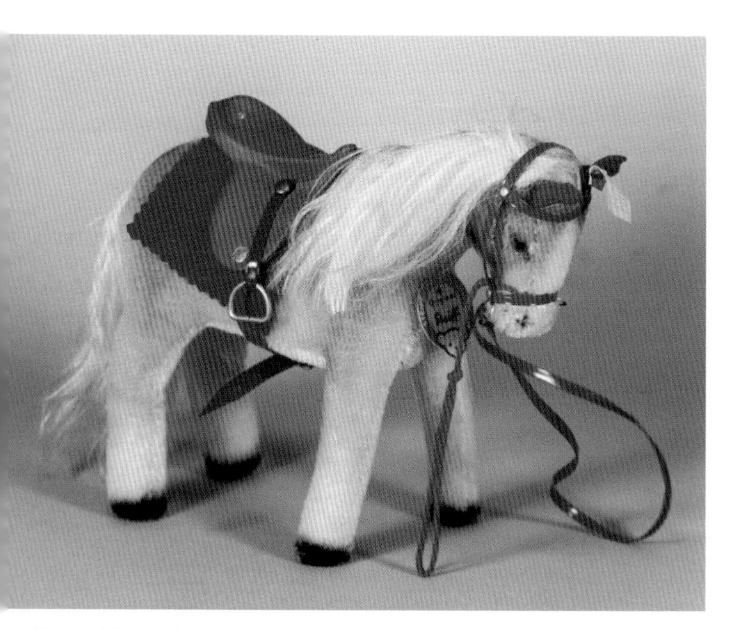

Above: Horse: 8 inches. Mohair mane swept to one side; glass eyes; felt blanket with molded plastic saddle; plastic and cotton cord trappings; raised script button; chest tag; 1950s. $500 up

Right: Horse on metal frame: 36 inches. Mohair; open felt mouth and inner ears; glass eyes; horsehair mane and tail; leather hooves; felt blanket. This is the smaller version of the life size model used in the 1958 Giengen, Germany, parade to celebrate Teddy Roosevelt's 100th birthday. $2,000 up

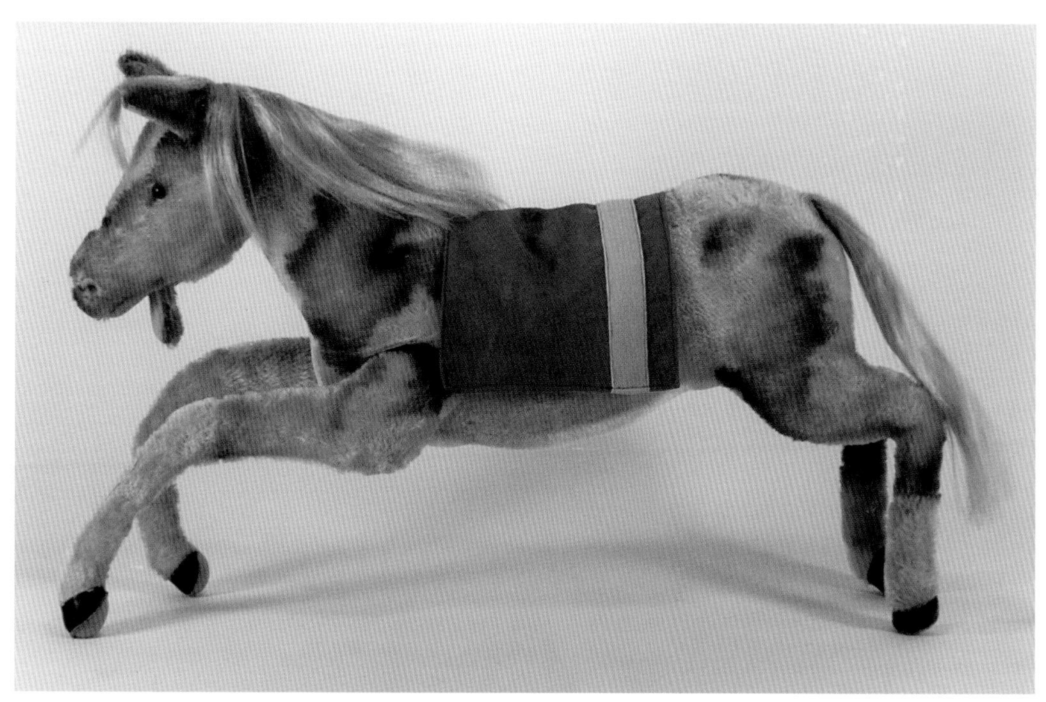

Horse: 12 inches. Airbrushed mohair; open felt mouth and nostrils; leather hooves; horsehair mane and tail; originally pulled a wagon; replaced blanket; offered by F. A. O. Schwarz in the 1950s; no I. D. $500 up

Ferdy Horse: 10 inches. Brown and white mohair; glass eyes; red leather bridle; circa 1957; raised button; bear head chest tag. $175

Horse: 9.50 inches. Acrylic velour; plush mane and tail; plastic eyes; soft but firmly stuffed; vinyl halter and reins; circa 1975; incised button; split chest tag; materials tag. $145

Cosy Pony: 9 inches. Brown and white plush; plastic eyes; excelsior and wool stuffed; cotton cord bridle; 1970s; incised button; split chest tag. $100

Sheddy Horse: 10 inches. Brown and white mohair; plastic eyes and bridle; circa 1973; incised button; split chest tag. $190-$195
Sheddy Horse: 7 inches. Same description as large. Incised button; split chest tag. $145-$155

Derby Horse: 9 inches. Beige plush with dark brown mane, hooves and tail; plastic eyes; circa 1985; brass button; split chest tag. $195

Young Donkey: 9 inches. Tan wool plush; mohair head and ears; glass eyes; felt hooves; printed FF button; 1930s. $295

Donkey: 5 inches. Gray velvet; glass eyes; leather bridle; circa 1960; raised script button; bear head chest tag. $145

Army Mule: 5 inches. Mohair; glass googly eyes; yarn bridle; felt blanket with gold A; mascot for Army Academy; 1950s; raised script button; split chest tag. $495

Grissy Donkey: 17 inches. Gray airbrushed dralon; open mouth; black glass eyes; circa 1960; raised button. $200 up

Grissy on rockers: 22 inches (plus rockers). Gray plush; glass eyes; felt open mouth; metal rockers; 1960s; raised button. $475

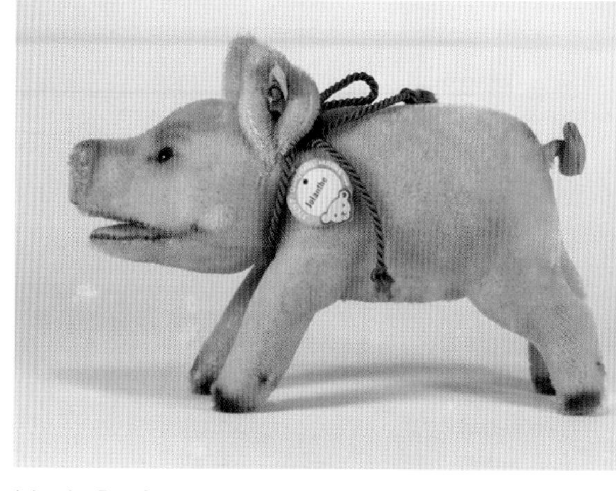

Jolanthe Pig: 9 inches. Pink mohair; blue glass eyes; open felt mouth and curled tail; green silk cord; one of several sizes; the smallest is velvet; circa 1955; raised script button; bear head chest tag. $250

Pig: 14 inches. Mohair; shoe button eyes; gauze nose with painted nostrils; side pushed oinker; circa 1929; printed FF button. $900

Lamby Lamb: 4 inches. Black and white curly plush; felt ears; green glass eyes; original bow and bell; resembles black lamb *Swapl*; in the absence of a chest tag the white tail and green eyes will identify this animal (*Swapl* has a black tail and blue eyes); 1950s; raised button; chest tag. $245

Lying Lamb: 10 inches. White wooly plush; green glass eyes; circa 1955; raised script button. $350

Lamby: 8 inches. White wooly plush; green plastic eyes; original ribbon and bell; circa 1955; raised button; chest tag. $175

Lamby: 12 inches. White wooly plush; glass eyes; excelsior stuffed; original ribbon and bell; circa 1965; incised button; chest tag; F. A. O. Schwarz tag. $245

Floppy Lamby: 15 inches. Wool plush; sponge stuffed; embroidered sleep features; original ribbon; circa 1970; incised button; chest tag. $125

Wotan Lamb: 9 inches. Wooly plush; plastic eyes; double weight felt horns; excelsior stuffed; circa 1955; raised button; chest tag. $395 up

Cosy Lamby: White and yellow airbrushed plush; open mouth; soft and hard stuffed; plastic eyes; circa 1965; incised button; chest tag; materials hang tag. $135

Floppy Lambs: 12 inches and 15 inches. Off-white plush; plastic eyes; soft stuffed; 1980s; brass buttons; split chest tags. Small $100; Large $160

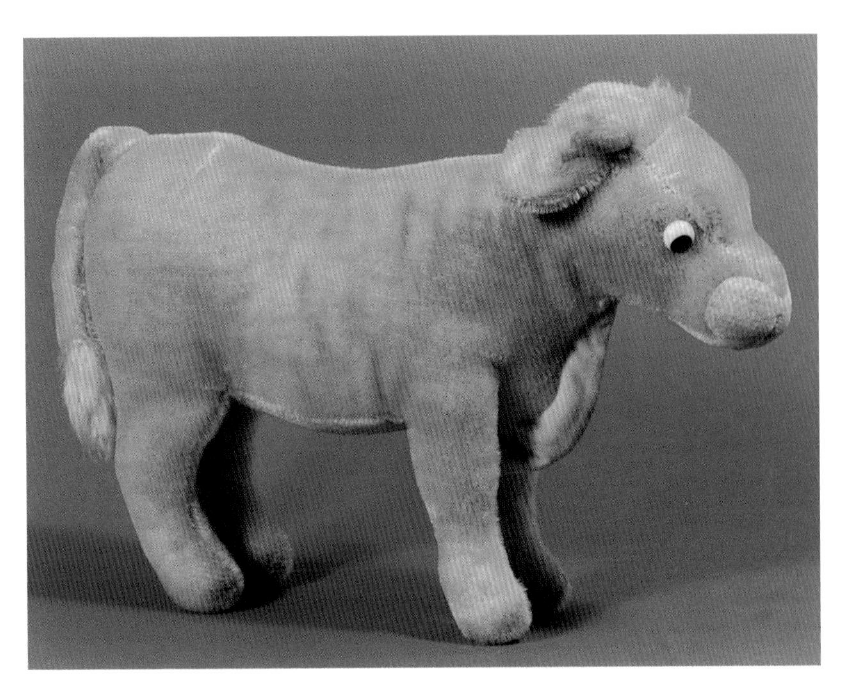

Oxy Oxen: 5 inches. Tan mohair; velvet nose and front piece; black and white glass eyes; 1950s; raised button. $250 up

Snucki Ram: 8.50 inches. Long white mohair; black velvet face, ears and legs; plastic eyes; felt horns; 1950s; raised button; chest tag. $150 up

Bessy Cow: 11 inches. Trevira velvet; soft stuffed; plastic eyes; circa 1970; incised button; split chest tag. $145

Cosy Flora Cow: 11 inches. Tan white and brown plush; plastic eyes; soft stuffed; collar and bell; 1980s; brass button; split chest tag. $95-$110
Flora Cow: 10 inches. Trevira velvet; plastic eyes; cotton and foam stuffed; 1978 only; incised button; split chest tag. $250 up

Goat: 6 inches. Mohair; glass eyes; circa 1930; printed FF button. $800 up

Goat: 9.50 inches. Off-white and black mohair; glass eyes; original ribbon and bell; circa 1930; printed FF button; orange stock tag. $1,800-$1,900

Rocky Goat: 7 inches. Tan mohair; wire enclosed felt ruled horns; green plastic eyes; 1950s; raised button. $85
Rocky: 5.50 inches. Tan mohair; doubleweight felt horns; green plastic eyes; 1950s; raised button; chest tag. $95
Zicky: 4 inches. Tan mohair; felt ears; green plastic eyes; ribbon and bell; 1950s; raised button; chest tag. $90

Zicky Goat: 4 inches. Mohair; glass eyes; felt ears; excelsior stuffed; original ribbon and bell; circa 1950; raised button; chest tag. $350-$400

Cosy Zicky Goat: 9 inches. Black and white dralon plush; glass eyes; felt horns; ribbon trim; circa 1960; raised button; chest tag; materials tag. $190

Above:: *Rocky* Goat: 8 inches. Airbrushed plush; felt ruled horns; plastic eyes; circa 1985; brass button; split chest tag; materials tag. $140

Right: Navy Goat: 9 inches. Long and short tan mohair; wired, padded and ruled horns; black and white glass eyes; felt blanket (the N has been replaced;) one of a series of college animals; also came in 6 inch size; 1950s; Chest tag. $325 up.

CHAPTER VIII

HIGH FLYERS & FOWL PLAY

STEIFF BIRDS ARE NEVER IN GILDED CAGES. THEY ARE FREE TO FLY INTO YOUR LIFE.

The first catalog, in 1892, shows birds of many descriptions. Under the heading of Vogel (bird), the illustration is of a duck sitting on a nest of eggs. The drawing is in black and white, but one can easily imagine the colorful display it presented. Among the feathered friends offered, besides ducks, were storks, chickens, canaries and parrots.

I find the vintage roosters particularly delightful with their multi-layered bits of colored felt meticulously applied. When the fabric of choice turned to mohair, it was beautifully airbrushed so the colorful presentation was retained. Once again, the variety of birds and fowl seems endless.

Birds: 7 inches, 4 inches and 3.50 inches. Mohair birds with swivel heads and metal legs; felt wings. Birds in foreground have wrapped wire legs and one leg moves the alternate wing; horsehair wing and tail tips; circa 1958; raised button; chest tags. Wing/tail birds $275-$350; Others $200-$275

Finch: 4 inches. Colorfully airbrushed mohair; felt tail; plastic feet, eyes and beak; circa 1960; raised button; chest tag. $125-$150
Finch: 4 inches. Colorful mohair; glass eyes; horsetail tail and wing tips; wire legs encased in vinyl; move wing by rotating alternate leg; circa 1960; raised button. $275-$350

Blackbird Mobile: 3 inches each. Three wool birds; paper wings; strung together; in cello box; circa 1970; incised button. $195 up

Wooden bird house with wool ball bird: 3 inches; paper logo on roof of the house; 1950s-1960s. $145

Wool Ball Birds: 2 inches and 1.50 inches. Two of several sizes and colors; wool yarn; plastic eyes, beaks and feet; incised buttons; stock tags; 1960. $25-$40

Hucky Crow: 7 inches. Black mohair; felt beak, tail and eye backs; glass eyes;metal feet; swivel head; raised button; chest tag. $145

Owl: 5 inches. Shaded blue mohair; plastic eyes; made for company in Germany; 1950s; raised script button. $225

Wittie Owl: 8 inches. Airbrushed mohair; felt feet and wing tips; horsehair tufts; swivel head; circa 1958; raised button; bear head chest tag. $125
Wooly Owl: 3.50 inches. Wool ball; felt nose; swivel head; plastic eyes and feet; circa 1965; incised button. $35-$50

Wittie Owl: 12 inches. Shaded mohair; horsehair ear tufts; large green glass eyes; rubber nose; unlike smaller sizes this has mohair feet; 1950s; raised button; chest tag. $295 up

Pigeon: 8 inches. Gray mohair; stiffened velour type wing tips and tail; plastic eyes (backed by felt,) beak and feet; circa 1955; incised button; chest tag. $195 up

Piccy Pelican: 10 inches. Tan with pinky airbrushed mohair; felt beak and feet; vinyl teeth; glass googly eyes; 1959-1961; raised script button; stock tag. $325

Parrot Mobile: 2 inches (each bird). 5 colorful wool parrots in plastic pack with dowels and plastic line to assemble; instructions with picture; mint in pack; 1960s; incised button on each bird. Box is fastened with 3 raised buttons. $250 up

Lora Parrot: 9 inches. Colorful mohair; plastic eyes; vinyl beak; circa 1968; raised button; chest tag. $150-$160

Franzi and *Hansi* Parakeets: 5 inches. Blue or green trevira velvet with felt tail; plastic eyes, beak and feet; circa 1973. $125 each. Shown with Steiff store sign.

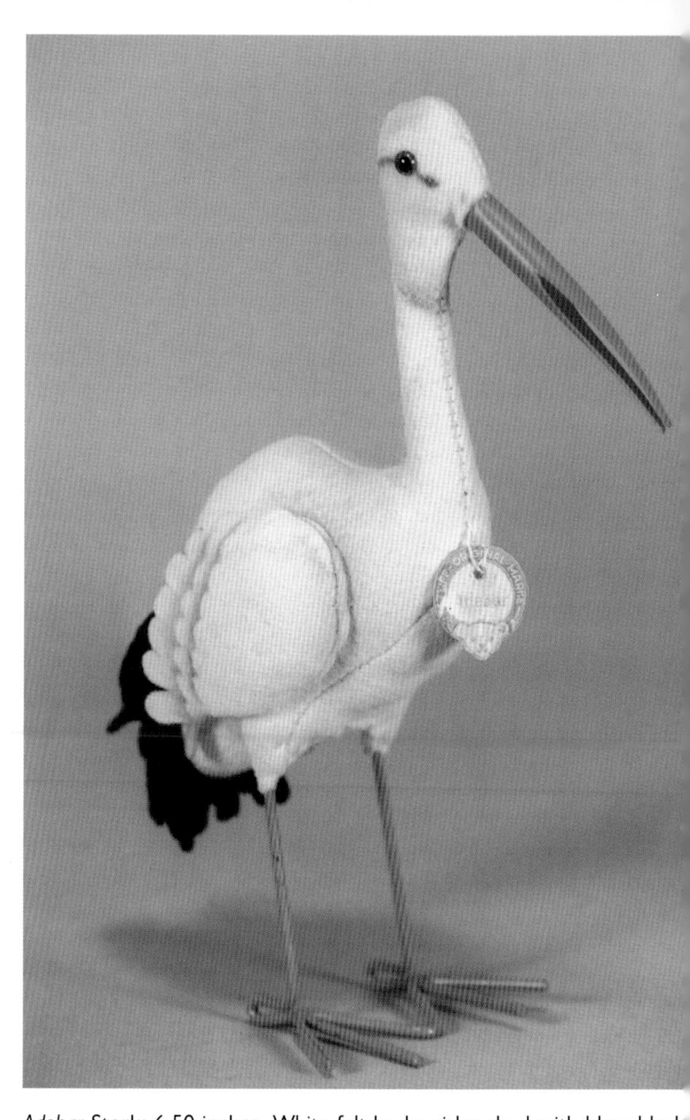

Adebar Stork: 6.50 inches. White felt body; airbrushed with blue; black tail feathers; plastic beak; metal legs; brown glass eyes; 1950s; raised button; chest tag. $225

Adebar Stork: 12 inches. Felt; glass eyes; airbrushed markings; metal armature in legs and beak; some moth damage to beak; circa 1950; no I. D. $450 up

Tucky Tucan: 12 inches. Colorful plush; plastic eyes; 1980s; brass button; split chest tag. $225-$245
Peli Pelican: 11 inches. Plush; plastic eyes; 1980s; brass button; split chest tag. $195-$225
Hucky Bird: 8 inches. Plush; plastic eyes; 1980s; brass button; split chest tag. $145-$155

Hen: 6 inches. Multicolored felt; wrapped wire feet; black bead eyes; 1905; no I. D. $450

Rooster and Hen: 3 inches. Mohair; felt heads and tails; metal feet; chest tags; 1950s. $95 each

Rooster: 7 inches. Airbrushed mohair; felt tail, comb and wattle; wired and padded felt feet; 1960s. $125
Hen: 6 inches. Same description as rooster except color is different; 1960s; incised button; chest tag on each. $125

Wool Chicken: 4.50 inches. Wooly plush; plastic eyes, beak and feet; 1970s; incised button; split chest tag. Shown with Steiff box of the period. $65

Wool Ball Chickens: 1 inch, 2 inch and 3 inch. Swivel heads; plastic eyes, feet and beaks; felt combs and tails on rooster and white hen; circa 1965; incised buttons. $35-$45 each

Chickens: 5 inches. Colorfully airbrushed plush; felt combs, wattles and tail feathers; plastic feet, beaks and googly eyes; 1980s; brass buttons; split chest tags. $75 each

Duck: 12 inches. Mohair; felt beak and feet; swivel head and legs; glass eyes; circa 1913. $500 up

Wooly Chicken: 4 inches. Plush; plastic beak, feet and eyes; swivel head; 1960. $75

Duck: 9 inches. Mohair; felt beak and wired feet; black glass eyes backed by red felt; circa 1920; no I. D. $250

Duck: 10 inches. Colorful mohair; felt beak and feet; in swimming position; glass eyes backed by red felt; circa 1920s; no I. D. $550

Tulla Goose: 6.75 inches. White mohair; felt beak and feet; plastic eyes; 1950s; raised button; chest tag. $165
Tulla Goose: 5 inches. Description the same; raised button; chest tag. $140
Ball body Chicken: 5 inches. Yellow mohair; straight out felt feet; plastic beak and eyes; 1950s; raised button; chest tag. $95
Ball body Chicken: 4 inches. Same description. $75

Wool Ball Duck: 3 inches. Yellow wool body; metal legs; glass eyes backed by felt; felt beak, hat and slippers; swivel head; 1930s. No I. D. $395

Kuschi Duck: 5 inches. Shaded off-white mohair; orange felt beak and feet; black glass eyes; chicken feather head tuft and tail; circa 1950; raised script button; chest tag. $500

Tulla Goose: 11 inches. Airbrushed mohair; felt beak and feet; glass eyes backed by felt; circa 1955; bear head chest tag. $695

Starli Drake: 15 inches. Beautifully and naturally colored dralon; plastic eyes; felt beak; ribbons on tail; circa 1975; incised button; chest tag. $350

Putty Turkey: 5 inches. Colorful plush; felt head, tail and wings; plastic eyes and feet; circa 1981; brass button; split chest tag; hang tag. $225

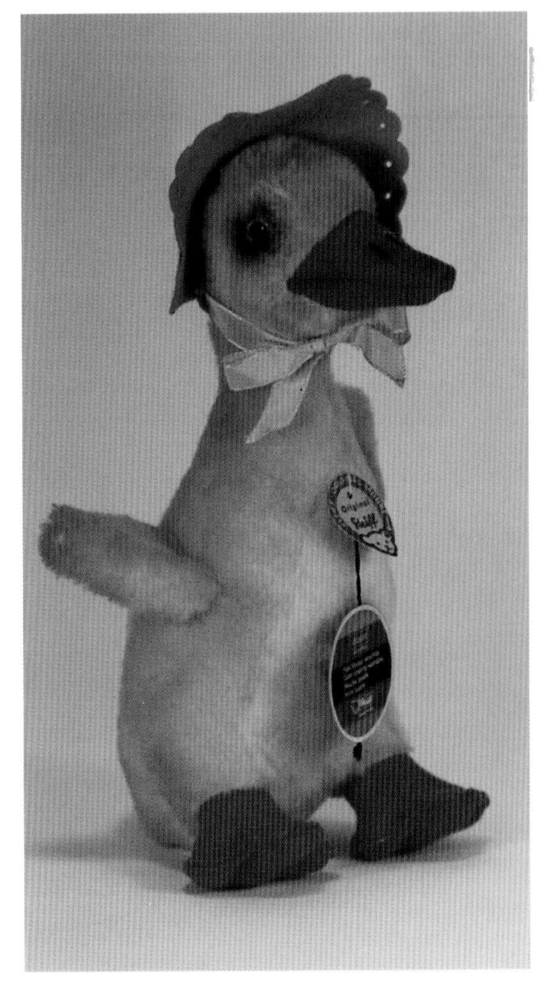

Duck: 10 inches. Wooly plush; plastic googly eyes; felt beak, feet and removable bonnet; circa 1960; incised button; bear head chest tag. $120

CHAPTER IX

DEER

DOES ARE DEERS, THEY'RE FEMALE DEERS.

I sit near my kitchen window in the early morning hours writing (or sometimes *thinking* about writing!) and am caught by the panorama of woodland drama. The deer come and nibble at the foliage or taste the water in the stream. It is a sight I never tire of. I also find I will never tire of or become blasé about Steiff's remarkable deer family. The company' s designers have caught the transformation from fawn to majestic buck with realism and perfection.

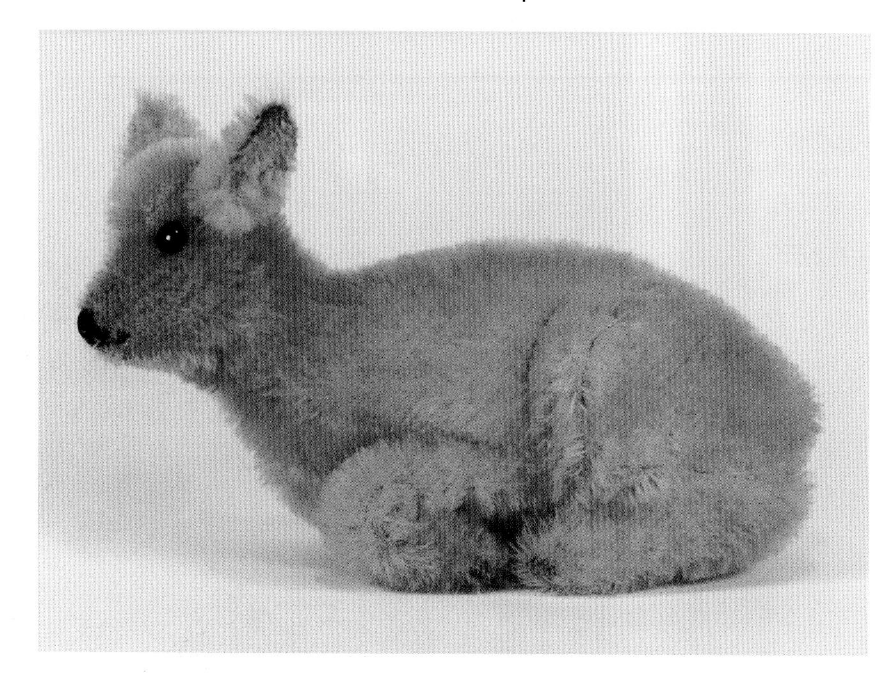

Fawn: 7.50 inches. Mohair; glass eyes; circa 1930; printed FF button. $700

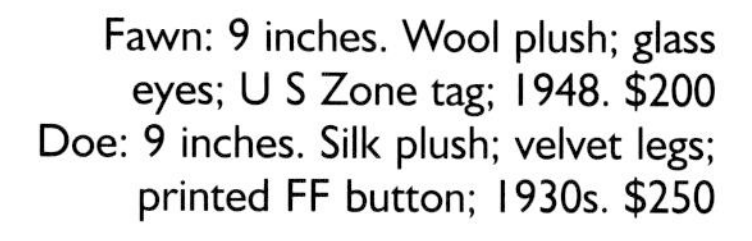

Fawn: 9 inches. Wool plush; glass eyes; U S Zone tag; 1948. $200
Doe: 9 inches. Silk plush; velvet legs; printed FF button; 1930s. $250

Doe: 12 inches. Airbrushed golden tan mohair; glass eyes; note forward head conformation typical of Steiff's female deer; circa 1950; raised button. $195

Buck: 12 inches. Fuzzy plush; felt covered wire antlers; glass eyes; circa 1948; no I. D. $225

Buck: 12 inches. Mohair; felt covered wire antlers; glass eyes; raised script button; stock tag; 1950s. $250

Doe: 9 inches. Tan mohair; plastic eyes; circa 1972; incised button; split chest tag. $140-$145
Fawn: 6 inches. Trevira velvet; plastic eyes; circa 1985; brass button; split chest tag. $95-$100

Renny Reindeer: 8 inches. Mohair; long mohair ruff; felt covered wire antlers; glass eyes; 1950s. $225

Above: Lying Fawn: 16 inches. Spotted mohair; glass eyes; note stitch on nose; circa 1957; raised button; chest tag. $350

Left: *Elfi* Deer: 16 inches. Airbrushed plush; plastic eyes; circa 1985; brass button; split chest tag; materials tag. $225-$235

CHAPTER X

IN THE GARDEN & WOODS

ON THE WAY TO GRANDMOTHER'S HOUSE WE GO OVER THE RIVER AND THROUGH THE WOODS ENCOUNTERING STEIFF CREATURES GREAT AND SMALL.

From the lowly field mouse *Pieps* to the exotic bat *Eric*, the Steiff range encompasses most animals found in the great out-doors. In 1897 a sitting squirrel and in 1913 a jointed fox were added to the line, but most garden and wood varieties were introduced around 1920. The greatest preponderance can be found in the post-World War II period, when their output and export reached its hey-day.

Squirrel: 9 inches. Off-white mohair; shoe button eyes; all jointed, including the tail; circa 1920. No I. D. $225 up

Hörni Squirrel: 13 inches. Plush; plastic eyes and nose; open felt mouth and teeth; cotton clothes; felt hat; dressed as a crossing guard; made for German bank; circa 1975; brass button; hang tag. $155

Perri Squirrels: 4 inches and 5 inches. Brown airbrushed mohair; glass eyes; felt hands and feet; large one holds acorn; circa 1955; raised buttons; chest tags. $110-$140
Possy Squirrels: 3.50 inches and 6 inches. Apricot mohair with gray tails; plastic eyes; large one holds velvet nut; circa 1960; raised buttons; chest tags. $110-$140

Fox: 10 inches plus tail. Realistic colored mohair; glass eyes; all jointed; 1920s; printed FF button. $795 up

Foxes: 5 inches and 7 inches. Rust, white and black mohair; larger version has longer mohair and a swivel head; glass eyes; circa 1950; raised script button. $145-$225

Chippy Ground Squirrel: 8.50 inches. Colorful airbrushed dralon plush; plastic eyes; circa 1975; incised button; split chest tag; materials tag. $140-$145

Sitting *Xorry* Fox: 6 inches. Orange and white airbrushed mohair; black tipped ears; glass eyes; circa 1955; note large ears; raised script button. $155 up

Cosy Fuzzy Fox: 8 inches. White and orange plush; plastic eyes; circa 1965; incised button; bear head chest tag; dralon hang tag; materials tag. $125

Xorry Fox: 4 inches. Orange and white mohair; glass eyes; velvet back of ears; circa 1958; raised script button; chest tag. $135 up

Xorry Fox: 12 inches. Airbrushed mohair; glass eyes; circa 1950; raised script button; chest tag; F. A. O. Schwarz tag. $395

Topsy Fox: 14 inches. Plush; plastic eyes; open mouth; Made for German bank; comes with book called *Topsy and Some Friends*. Circa 1980; brass button. $175

Wiggi and Waggi Weasels: 8 inches. White (*Wiggi*) or white and brown (*Waggi*) plush; pipe cleaner tails; black glass eyes; felt feet and ears; circa 1960; raised script buttons. $150 up

Minky Weasel: 11 inches. Plush; felt ears and feet; glass eyes; 1960s. $90

Steiff Wool Ball Bird: 2 inches. Wool ball body and head; felt tail; plastic feet; swivel head; circa 1960; no I. D. $30-$40
Wool Ball Lady Bugs: 1.50 inches and 2 inches. Airbrushed with bead eyes, antennae, and swivel heads; circa 1960. No I. D. $34-$40

Eric Bat: 9 inch (wingspan). Mohair; felt ears; pipe cleaner limbs; glass bead eyes and nose; vinyl wings; circa 1958; raised button; bear head chest tag. $295-$325

Phuy Skunk: 8 inches. Mohair; glass eyes; felt feet; chest tag; 1950s. $350 up

Skunk: 4 inches (plus tail.) Mohair and velvet; glass eyes; circa 1955; raised button; bear head chest tag. $225

Spidy: 8 inches. Mohair; wired legs; 7 glass bead eyes; 1950s; raised script button. $375

Lizzy Lizard: 12 inches. Velvet; felt feet; glass eyes; circa 1958; raised button; chest tag. $245 up

Swinny Hamster: 4.50 inches. White, black and orange plush; felt feet; plastic eyes; circa 1959; raised script button; bear head chest tag. $85-$95

Diggy Badgers: 4 inches and 6 inches. White, black and pale orange mohair; double weight felt paws; glass eyes; circa 1950. Chest tag on larger; all I. D. on smaller. $140-$145

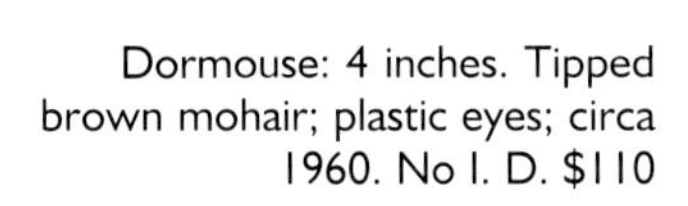

Dormouse: 4 inches. Tipped brown mohair; plastic eyes; circa 1960. No I. D. $110

Piff Groundhog: 7 inches. Shaded mohair; open felt mouth; double felt hands; stuffed felt feet; glass eyes; circa 1955; chest tag. $145
Piff Groundhog: 4 inches. Same description as larger except both hands and feet are double felt; chest tag. $110

Mouse: 5 inches. Mohair with felt ears and paws; rope tail; plastic eyes; one of the animals from *"Alice And Her Friends;"* boxed set of the 1980s; brass button. $95

Pieps Mouse: 3 inches. Mohair; felt feet and ears; pink glass eyes; black glass nose; all I. D. 1950s. $95 up

CHAPTER XI
RABBITS

LOOK WHO'S HOPPING DOWN THE BUNNY TRAIL. IT'S BUNNIES BY THE SCORE.

Next to cats and dogs, rabbits appear to be the most desirable animal in the adult marketplace. One sometimes tends to forget that Steiff products were originally meant for children, but fortunately they like bunnies too, so the collector has no problem seeing his collection multiply.

Rabbits have been made in great quantities since 1900 because, for one thing, they were and are the obvious choice for inclusion in Easter baskets. One of the most interesting, and one of the most difficult to find, is the *Holland* rabbit. Besides being jointed five ways (head and four legs), the ears were also articulated and thus could move in a natural way.

There have been many fascinating rabbit products over the years, so their popularity is not surprising. *Peter Rabbit*, produced for the English market in 1905, emulated Beatrix Potter's tale and wears a felt jacket and slippers. Another character from children's literature was *Jack Rabbit* (also dressed). Jack was fashioned in two sizes from 1927 to 1931. Because only 2780 examples were manufactured, it is easy to fathom the rarity value.

The 1950s brought a multitude of other designs. *Manni,* made in sizes from 3 ½ inches up, has proven to be a perennial favorite. Another rarity, and one I have only seen a picture of, is a jointed rabbit much like the same decade's *Niki,* except he sports tremendous ears. He is named *Turnip Rabbit*, but actually holds a large orange carrot. Whatever one desires, Steiff seems to have satisfied aficionados wants when it comes to the bunny brigade.

Rabbits: 3.50 inches and 4 inches. Velvet with airbrushed spots; bead eyes; original ribbons and bells; some minor foxing; circa 1905; blank buttons. $800-$1,000

Peter Rabbit: 4 inches to 12 inches. Produced from 1905 to about 1915; velvet; felt coat and slippers. Photo courtesy of Sue and Michael Pearson. $2,000 up

Holland Rabbit: 13 inches (plus ears). Mohair; pink inner ears; pink glass eyes; head, legs and ears jointed; printed FF button; 1908. $1,600 up

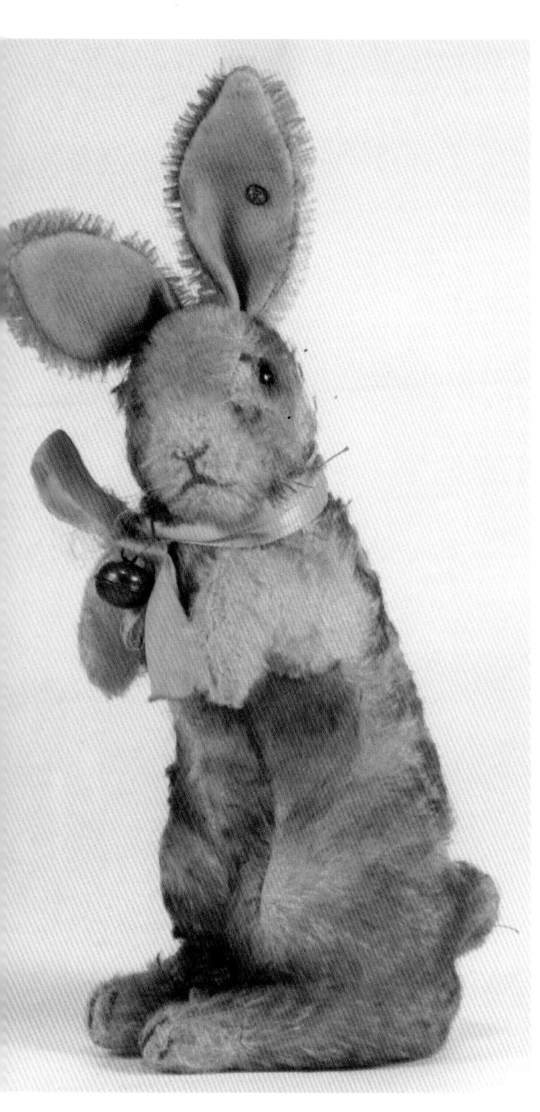

Rabbit: 8 inches (plus ears). Cream mohair with brown tipping; glass eyes; circa 1930. Printed FF button. $850 up

Rabbit: 9 inches. Off-white mohair; glass eyes; shows wear; no I. D. Circa 1925. $250

Tail/Head Rabbit: 6 inches. Off-white mohair; glass eyes; head moves in a circular motion by tail leverage; no I. D. Circa 1930. $800 up

Tail/Head Rabbit: 9 inches. Tan mohair; glass eyes; circa 1930; printed FF button. $1,050 up

Rabbit: 7 inches. Mohair; glass eyes; original ribbon and bell; circa 1948; U S Zone tag. $190-$195

Running Rabbit: 7 inches. Shaded mohair; glass eyes; original ribbon; circa 1955; raised button; chest tag; F. A. O. Schwarz tag. $155

Above: *Manni* Rabbit: 17 inches (plus ears). Airbrushed mohair; open felt mouth; large glass eyes; swivel head and arms; 1950s; raised script button. $775

Left: *Manni* Rabbit: 14 inches (plus ears). Tan shaded mohair; glass eyes; open felt mouth; swivel head; original ribbon; circa 1955; raised button; chest tag. $525-$535

Varlo Rabbit: 5 inches. Off-white mohair; glass eyes; jointed head and back legs; circa 1959; raised button; chest tag. $175

Easter Bunny: 9 inches (plus ears). Mohair; glass eyes; swivel head and arms; basket with grass on back; circa 1955; raised script button. $350

Niki Rabbit: 8 inches. Mohair; glass eyes; open felt mouth; all jointed; original ribbon; circa 1957; raised button. $250
Lying Rabbit: 6 inches. Tan mohair; glass googly eyes; original ribbon; circa 1958; raised button. $120

Niki Rabbit: 12 inches (plus ears). Mohair; large felt feet and open mouth; glass eyes; all jointed; circa 1955; raised button. $400

Cosy Mummy: 8 inches. Wooly plush; glass eyes; soft stuffed; circa 1959; raised button; bear head chest tag. $75-$80. *Timmy* Rabbits: 4.50 inches and 6 inches. Black and white wooly plush; glass eyes; circa 1960; incised buttons; bear head chest tags. $100 and $120

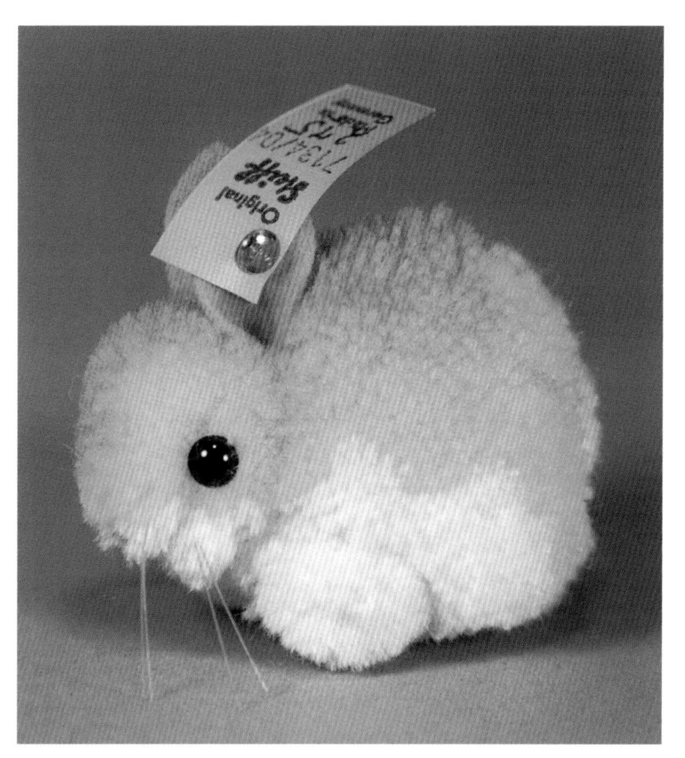

Above: Wool Ball Rabbit: 2 inches. Plastic eyes; felt ears; swivel head; one of several colors; 1960s; incised button. $25-$30

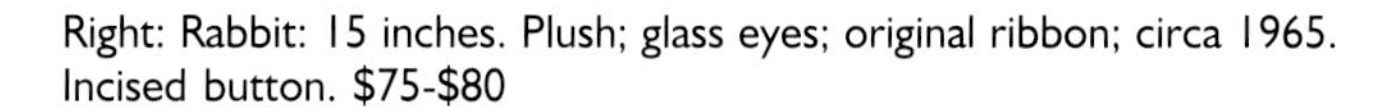

Right: Rabbit: 15 inches. Plush; glass eyes; original ribbon; circa 1965. Incised button. $75-$80

Ossi Rabbit: 9 inches. White plush and brown tipped mohair; glass eyes; ribbon trim; 1960s; incised button; bear head chest tag. $150
Pummy Rabbit: 5 inches. Mohair; plush inner ears, chest and tail; swivel head; cotton stuffed; plastic eyes; 1960s; no I. D. $95

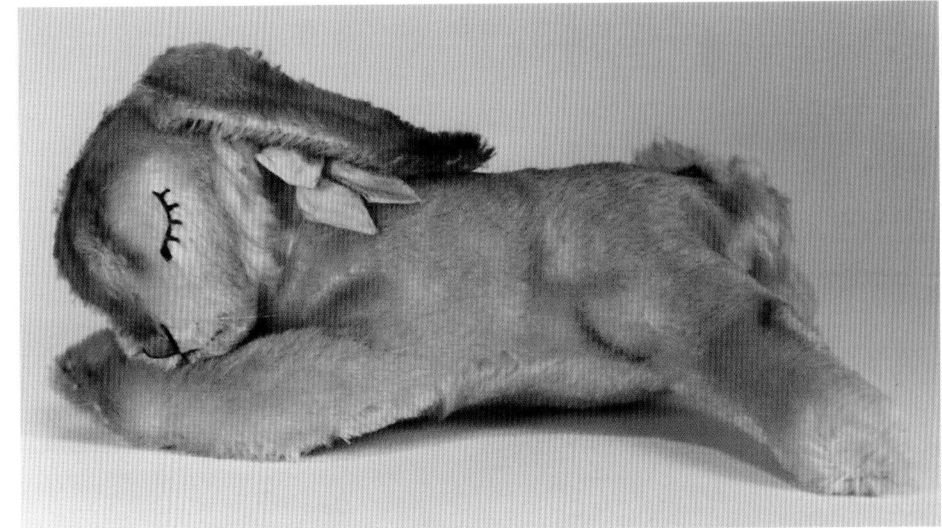

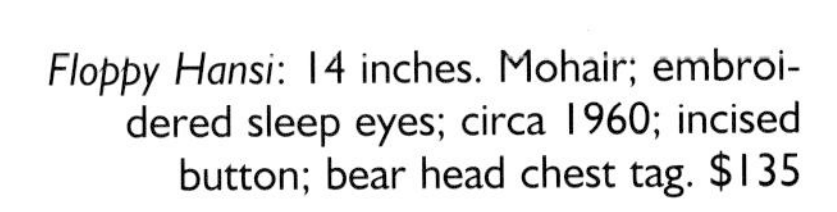

Floppy Hansi: 14 inches. Mohair; embroidered sleep eyes; circa 1960; incised button; bear head chest tag. $135

Sonny Rabbits: 6 inches and 8 inches. Mohair; glass eyes; swivel heads; 1960s; incised buttons; bear head chest tags. $145 and $165

Sitting Rabbit: 8 inches. Wooly plush; plastic eyes; swivel head; circa 1960; incised button; bear head chest tag. $125

Lausy Rabbit: 12 inches. Plush; plastic eyes; circa 1975; incised button; split chest tag. $95-$100
Snuffy Rabbit: 7 inches (sitting). Plush; blue plastic eyes; circa 1975; incised button; split chest tag. $50

Dürer Style Rabbit: 11 inches. Tipped and wooly plush; plastic eyes; 1980s; brass button; split chest tag; hang tag. $135-$150
Baby Rabbit: 12 inches. Plush; made-on suit with cord trim; plastic eyes; 1970s; incised button; split chest tag. $85

Dormy Rabbit: 10 inches. Tan natural looking plush; crouched position; circa 1985; brass button; split chest tag. $85-$95
Snobby Rabbit: 15 inches. Brown plush; plastic eyes; lying position; circa 1985; brass button; split chest tag. $85-$95
Spotty Rabbit: 12 inches. Black and white plush; plastic eyes; lying position; circa 1985; brass button; split chest tag. $140-$150

Dormy Rabbit: 13 inches. Plush; plastic eyes; circa 1980; brass button; split chest tag. $170-$180
Mummel Rabbit: 9 inches. Plush; plastic eyes; circa 1975; incised button; split chest tag. $70-$75

Lulac Rabbits: 11, 20 and 26 inches. Plush; plastic eyes; open felt mouths; long dangly legs. The two largest are from the 1970s; smaller 1980s. (*Lulac* has been made since the 1960s.) 11 inch brass button; split chest tag. 20 and 26 inch incised buttons; split chest tags. 11 inch $60-$65; 20 inch $225-$235; 26 inch $315-$345

CHAPTER XII

ELEPHANTS

IN OKLAHOMA THEY SAY THE CORN GROWS AS HIGH AS AN ELEPHANT'S EYE.

Elephants have a special significance in the Steiff Company. Although the camel was actually the first *registered* trademark, that bit of history seems to have fallen by the wayside and is barely acknowledged.

The first toy, released by the founder's nimble fingers, was an elephant and this fact is obviously noteworthy. From 1897 until sometime in the third decade of the twentieth century, when the Teddy Bear took precedence, the elephant became the company's logo. The first paper label used in 1897 and 1898 was square, attached by string and showed the mighty beast with trunk upraised. From 1900 to 1904 a round, thin cardboard tag appeared and also depicted an elephant with trunk upraised.

The first button Steiff identification mark, used in the left ear, was of silver colored metal and featured an embossed elephant. Since this was in use for a scant two years and is mostly found on rod bears and animals, a piece with this identification mark is highly coveted.

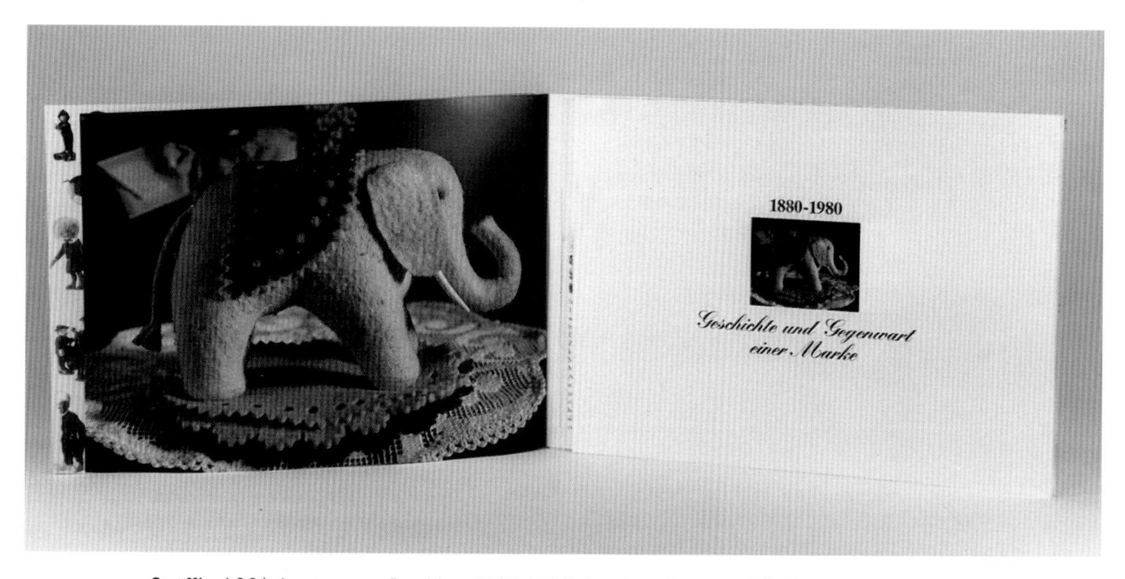

Steiff's 100th Anniversary Booklet: 1880-1980; interior photo on left shows the 1st product (an elephant pincushion). The cover is shown on the right side. $45

"Rod" Elephant: 16 inches. Gray wool coat, napped fabric; shoe button eyes; jointed by inner rods; circa 1904; no I. D. $2,800

Elephant: 8 inches. Curly cotton fabric; glass eyes backed by felt; original striped cotton blanket with pom poms; circa 1930; retains original price tag; printed FF button; orange stock tag. $1,200

Tail/Head Elephant: 14 inches. Pink mohair (faded from lavender;) glass eyes; airbrushed toes; felt trunk; the tail moves the head in a circular fashion; working squeaker; some mohair loss; printed FF button. $2,000 up

Circus Elephant: 32 inches. Gray mohair; black glass eyes; wooden tusks; raised script button; 1950s. $3,000

Elephant: 4 inches. Gray mohair; felt ears; rope tail; glass eyes; felt neck ornament; circa 1958; raised button. $95

Elephant: 15 inches. Tan mohair with black airbrushed feet; black and white glass eyes; felt tusks; long mohair tail tassel; red and yellow felt blanket; circa 1950; raised button. $550 up

Snuggy Jumbo: 31 inches. Mohair; glass eyes; felt tusks; suedene blanket; made especially for children to sit on while watching T V. 1950s. $1,200

Mini Elephant: 3 inches. Mohair; glass eyes; rope tail; felt pads and ears; suedene blanket with logo; circa 1950. $100

Jumbo Elephant: 13 inches sitting. Largest size of two; gray mohair with airbrushing; open felt mouth, inner trunk and foot pads; black and white googly eyes; red felt bib; circa 1950; raised button. $450

Jumbo Elephant: 9 inches. Gray mohair; swivel head and arms; plastic googly eyes; felt pads and open mouth; upraised trunk with bell on end; circa 1960; incised button; chest tag. $175

Elephant Head Trophy: 12 inches. Gray mohair; black glass eyes; open mouth with upraised trunk; felt tusks and trunk end; mounted on wooden board for hanging; circa 1955; raised button; stock tag; chest tag. $375 up

Cosy Trampy: 7 inches. Plush; velour pads; plastic eyes; collar with bell; incised button; chest tag; 1960s. $125

Baby Hathi Elephant: 9 inches. Dralon; made for the 2nd series of *The Jungle Book*; © Disney Enterprises. 1970s; brass button; split chest tag. $175-$185
Floppy Ele Elephant: 11 inches. Gray mohair; foam rubber stuffed; red felt blanket; circa 1972; incised button; split chest tag. $125-$130
Cosy Trampy Elephant: 8 inches. Airbrushed pink and gray dralon; plastic eyes; collar and bell; 1960s; button; chest tag. $125

Trevira Velvet Elephants: 6 inches and 10 inches. Plastic eyes; felt tusks; felt ears on smaller example; incised button on the largest; split chest tags on both; 1970s. $125-$175

Wooly Anniversary Elephant: 2 inches. Felt ears; bead eyes; felt blanket; carries scroll in trunk concerning 100th year; complete in plastic case. 1980; brass button. $125

Replica Elephant of the first piece made by Margarete Steiff. 1984-1987. $100

CHAPTER XIII
MONKEYSHINES

WHILE THE MONKEYS MAY HAVE HAD NO TAILS IN ZAMBOANGA, STEIFF'S DO AND WE CAN ENVISION THEM USING THIS APPENDAGE TO SWING FROM TREE TO TREE.

One of the rarest uses for a primate in the Steiff program was as a radiator cap designed in 1912/1913. The Steiff brothers were the first to own a vehicle in Giengen and this original advertising ploy was promptly implemented. The chimp model came nattily chapeaued in a chauffeur' s cap and held a brass steering wheel.

The company's primates have extraordinary appeal and have been produced since 1892. The many innovative examples include the early, string jointed circus clowns, roly polies, and even one smoking a cigar. Baboons, gorillas and orangutans are a few of the species, but the most recognizable is *Jocko* the Chimp, a long time favorite.

Monkey: 14 inches. Brown mohair; felt face, ears and paws; shoe button eyes; all jointed; shows some wear; circa 1910. No I. D. $495

Monkey: 10 inches. Brown tipped mohair; felt face with open mouth, ears and paws; glass eyes set into lids; head moves by rotating the tail; shows wear; circa 1930; printed FF button. $575-$600

Monkeys in cage: 5 inch *Jocko* monkeys with chest tags; 1950s; possibly assembled by F. A. O. Schwarz. $750 complete

Jocko: 19 inches. Brown mohair; felt face and paws; glass eyes; all jointed; circa 1955; chest tag. $450

Jocko Monkeys: 7 inches. White and brown mohair; green glass eyes; felt faces, paws and feet; all jointed; circa 1955; chest tags. $125 each

A five foot *Jocko* of the 1950s holds a smaller version. No I. D. $2,000 up

Coco Baboon: 6 inches. Gray mohair; tan felt face and ears; gray felt paws; red felt backside; green glass eyes; 1950s; raised button; chest tag. $165

Coco on eccentric wheels: 10 inches. Pale gray mohair; tan felt face and ears; green glass eyes; red felt hat; swivel head; wheels offset resulting in hopping action when pulled; 1948; raised button; trace of U S Zone tag. $425

Cocoli Bellhop: 10 inches. Tan, black and red felt; glass eyes; non-removable clothes; some moth damage; circa 1948. Full U S Zone tag. $450

Cosy Gora: 12 inches (arm span). Plush gorilla; plastic googly eyes; wears diapers; 1960s; raised script button; bear head chest tag. $295

Music Jocko: 11 inches. Brown mohair; felt face, paws and ears; all jointed; music box is activated by pressing felt circle (circle is replaced); circa 1955; raised script button. $1,250

Jocko Monkey: 25 inches. Brown curly mohair; felt face, ears, paws and feet; glass eyes; all jointed; circa 1955; raised script button. $695

Coco Baboon: 12 inches: Mohair; felt face with open mouth; glass eyes set into lids; horsehair mantle; all jointed; all I. D. 1950s. $650
Coco Baboon: 6 inches. Mohair; felt face, paws and ears; swivel head; all I. D. $165

Gibbon: 10.50 inches. Mohair; velour face and feet; plastic googly eyes; hands attached to head; 1960; raised script button; F. A. O. Schwarz tag. $195-$225

Hango Monkey: 6 inches. Shaded tan dralon; glass eyes; felt ears and paws; swivel head; 1960s; chest tag. $175

Mungo Monkey: 9 inches. Plush; velour hands and feet; glass eyes; 1960s; all I. D. $150

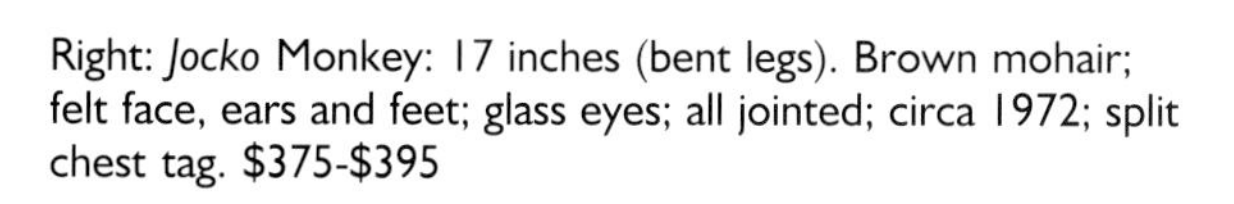

Above: *Mungo* Monkey: 9 inches. Colorful plush; velour face, hands and feet; plastic eyes; circa 1975; split chest tag. $95

Right: *Jocko* Monkey: 17 inches (bent legs). Brown mohair; felt face, ears and feet; glass eyes; all jointed; circa 1972; split chest tag. $375-$395

Kleideraffe Monkey: 22 inches. Plush; velour face, ears and paws; plastic eyes; 1980s; all I. D. $495

Mungo Monkey: 10 inches. Plush; plastic eyes; velour paws; swivel head; 1980s; brass button; split chest tag. $115-$125

Baby Orangutan: 13 inches. Rust, shaggy plush; velour face, hands and feet; plastic eyes; 1980s; brass button; split chest tag. $145

Gora Baby: 13 inches. Shaded, shaggy plush; velour face, hands and feet; plastic eyes; 1980s; brass button; split chest tag. $145-$155

CHAPTER XIV

IN AND OUT OF AFRICA

SIEGFRIED AND ROY ARE THE ONLY PEOPLE I KNOW OF WHO WOULD DARE TO HOLD THAT TIGER!

Wild animals roam from the plains of the American west to the African Serengeti. Every continent on earth teems with fauna striving to survive in their natural habitat. Many animals are endangered and we can only hope and do our bit to make sure, delightful though they are, that Steiff''s replicas are not the only way we can see them.

A stroll through the Safari and Game Preserve of Steiff is an education in itself. Many people had never heard of an okapi until they encountered the three sizes in the Steiff program. How clever to choose a little known African animal of unusual beauty and make it famous. Wander through the following pages and enjoy Australian koalas, the pandas of China, the llamas of South America, Indian tigers, the mighty beasts of Africa and all the rest of the world's strange, exotic and wonderful creatures.

Yuku Gazelle: 9 inches. Mohair; vinyl horns; glass eyes; circa 1958; raised button; chest tag. $350 up

Giraffe: 6 inches. Spotted velvet; glass eyes; felt ears; rope tail; circa 1960; incised button; chest tag. $150

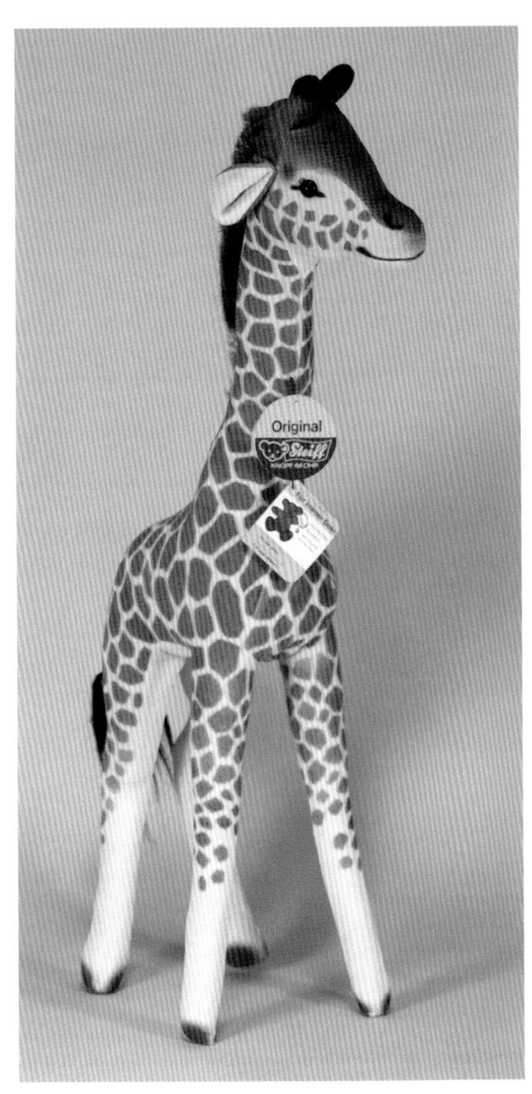

Giraffe: 24 inches. Synthetic covering; plastic eyes; 1980s; brass button; split chest tag; hang tag. $190-$195

Trevira velvet Giraffe: 10 inches. Spotted velvet; plastic eyes; circa 1972; brass button; chest tag; materials tag. $150
Trevira *Nosy* Hippo: 7 inches. Airbrushed velvet; plastic eyes; felt horns and ears; circa 1972; incised button; chest tag; materials tag. $125

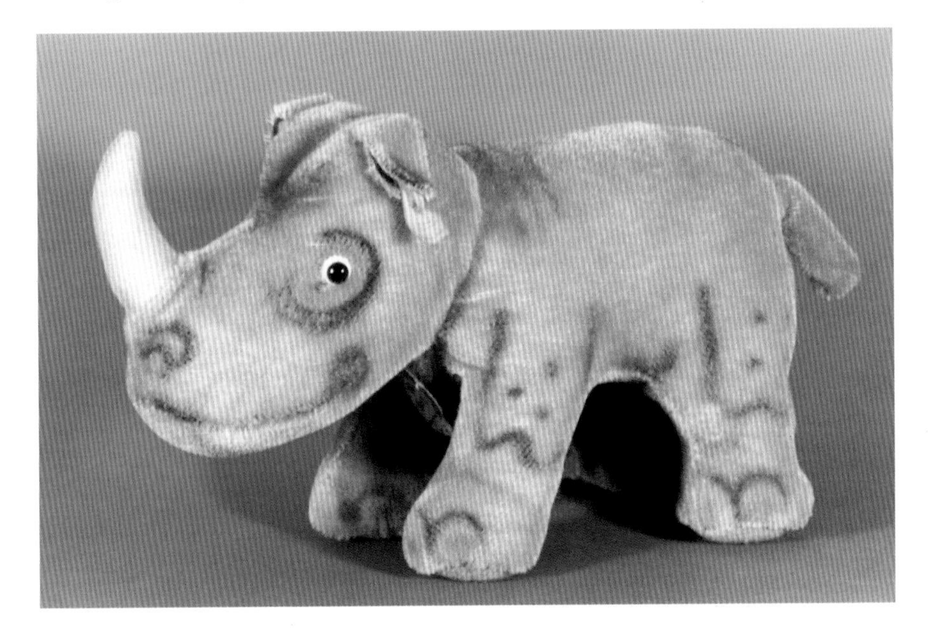

Rhino: 11 inches. Gray shaded mohair; felt horns; glass googly eyes; raised button; chest tag. $225

Mockie Hippo: 4.50 inches. Gray mohair; open felt mouth; glass googly eyes; rope tail; circa 1960; raised button; chest tag. $125
Rhino: 5 inches. Gray mohair; felt horn and ears; glass googly eyes; circa 1960; chest tag. $125

Nosy Rhino: 11 inches. Trevira velvet; plastic eyes; felt ears; circa 1975; incised button; split chest tag; hang tag. $135-$145

Mockie Hippo: 8 inches. Trevira velvet; plastic eyes; circa 1975; incised button; split chest tag. $125

Nase Vorn: 12 inches. Comic character rhinoceros made for the German market; acrylic with plastic eyes; 1980s; brass button; special large chest tag reads Aus der Fernseh-show/Nase Vorn/Von Frank Elster/Original Nasevornhörnchen/Von Steiff. $125

Rhino *Nase Vorn*: 12 inches. Royal blue and black plush; plastic eyes; one of several colors; designed from a character by Frank Elstner for the German market; 1980s; brass button; special chest tag. $125

Reclining Leopard: 20 inches. Realistically colored and spotted mohair; green glass eyes; 1950s; no I. D. $250

Leopard: 17 inches. Spotted plush; plastic eyes; circa 1985; brass button; split chest tag; materials tag. $275

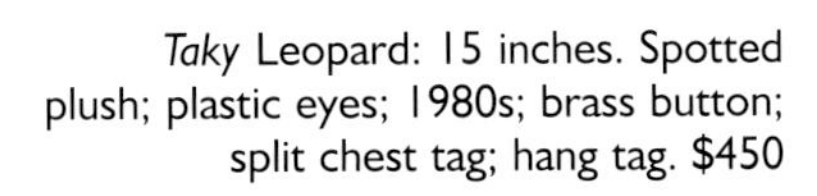

Taky Leopard: 15 inches. Spotted plush; plastic eyes; 1980s; brass button; split chest tag; hang tag. $450

Okapis: 6 inches and 16 inches. Smaller one is velvet; large one is mohair; glass eyes; 1950s; raised script button on the larger and chest tag on the smaller. $165 and $450 up

Okapi: 10 inches. Beautifully airbrushed velvet; mohair inside ears and tail ruff; glass eyes; circa 1958; raised button; chest tag. $295 up

Zebra: 8.50 inches. Black and white striped mohair; glass eyes; circa 1960; chest tag. $225
Zebra: 5 inches. Black and white striped velvet; rope tail; glass eyes; circa 1960; raised button; chest tag. $125

Zebra: 8 inches. Black and white striped mohair; plastic eyes; circa 1975; incised button; split chest tag. $175

Lioness: 10 inches. Gold mohair; glass eyes; circa 1949. No I. D. $275

Lion Cub: 8 inches. Wooly airbrushed plush; glass eyes; swivel head; circa 1948; raised button; U S Zone tag. $375

Lion Cub: 16 inches. Tan spotted mohair; glass eyes; 1950s; raised script button. $185

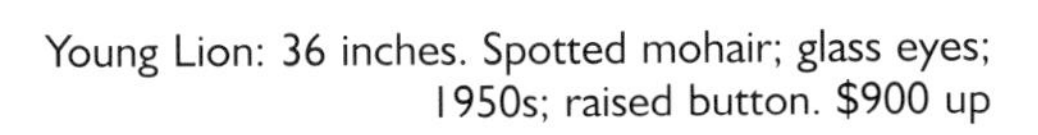

Young Lion: 36 inches. Spotted mohair; glass eyes; 1950s; raised button. $900 up

Lea Lioness: 4.50 inches. Airbrushed mohair; glass eyes; circa 1955; chest tag. $145
Tiger: 5 inches. Airbrushed striped mohair; glass eyes; circa 1955; raised script button; chest tag. $175

Sitting *Leo* Lions: 4.50 inches and 9 inches. Gold mohair; long tipped mohair manes; glass eyes; 1950s and 1960s production; large has incised button; small has raised button; both have bear head chest tags. Small: $145; Large: $175

Lion: 15 inches. Gold plush; plastic eyes; circa 1970; incised button. $95

Baby Lion: 17 inches (plus tail). Spotted plush; plastic eyes; circa 1975; split chest tag. $125 up

Lion in Cage Box: 7 inches. Plush; plastic eyes; circa 1972; comes with cage presentation box; incised button; split chest tag; booklet. $175

Tiger in Cage Box: 7 inches. Same description except has bear head chest tag; circa 1970. $175

Tiger: 8 inches. Airbrushed striped mohair; plastic eyes; 1974.

Leo Lion: 7 inches. Long and short mohair; plastic eyes; 1974; incised buttons; split chest tags. $125 each

Ocelot: 9 inches. Mohair; *yellow* eyes (no neck tufts like leopard); circa 1955. No I. D. $275

Luxy Lynx: 7 inches. Apricot shaded mohair; orange eyes; horsehair ear tufts; circa 1960. No I. D. $295

Cosy Puma: 13 inches. Very soft plush; plastic eyes; circa 1985; brass button; split chest tag; materials tag. $140

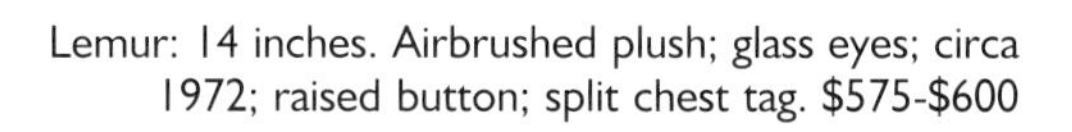

Lemur: 14 inches. Airbrushed plush; glass eyes; circa 1972; raised button; split chest tag. $575-$600

Llama: 11 inches. Long and short mohair; glass eyes; circa 1960; raised button; chest tag. $275

Llama: 16 inches. Long and short mohair; glass eyes; circa 1955; raised button; chest tag. $695

Moose: 7 inches and 11 inches. Mohair; glass eyes; felt antlers; 1950s; chest tag on larger, no I. D. on smaller. $275 and $600

Bison: 14 inches. Brown and gold airbrushed mohair; long mohair head, ears, mantle and tail; felt stuffed and ruled horns; glass eyes; chest tag; 1950s. $500

Loopy Wolf: 10 inches. Mohair; plastic eyes; circa 1958; all I. D. $1,800 up

Bison: 9 inches. Long and short mohair; glass eyes; felt horns; circa 1958. No I. D. $175

Wild Baby Boar: 5 inches. Tan spotted and striped mohair; bright blue glass eyes; felt ears and snout; rope tail; all I. D. 1950s. $150

Dalle Boar: 8 inches. Brown bristle mohair; brown glass eyes (earlier ones have blue); plastic tusks; incised button. 1960s. $75

Wild Boar *Dalle*: 11 inches. Mohair; felt on end of snout; glass eyes; plastic tusks; rope tail; 1950s. All I. D. $185

Koala: 5 inches. Shaded mohair; glass eyes; felt nose; stuffed felt feet; swivel head; circa 1955; chest tag. $395

Koala: 9 inches. Mohair; glass eyes; felt nose; all jointed; 1950s; no I. D. $450 up

Koala Bear: 15 inches. Two tone tan mohair; glass eyes; felt nose and open mouth; all jointed; circa 1950. No I. D. $700 up

Above: *Cosy Koala*: 7 inches. Brown and white dralon plush; velour airbrushed foot pads; plastic eyes; sitting position; swivel head; circa 1965; incised button; bear head chest tag; dralon hang tag. $95 up

Right: *Kangoo* Kangaroo: 6 inches. Mohair shaded; plastic eyes; plastic baby; 1950s; raised button; chest tag. $125

Kangoo Kangaroo and Baby: 20 inches. Mohair with jointed arms; glass eyes; velvet Joey is 4 inches; 1955; raised buttons on both; chest tag on *Kangoo*. $800 up

Linda Kangaroo: 10 inches. Airbrushed mohair; glass eyes; swivel head and arms; plastic baby; circa 1972. $195
Linda Kangaroo: 20 inches. Same description as smaller except baby is velvet; incised buttons; split chest tags. $595

Floppy Panda: 12 inches. Mohair; open felt mouth; embroidered sleep eyes; airbrushed pads; no I. D. 1960s. 10 inch Panda described elsewhere. Lorraine Oakley collection. $200

Panda Bear: 10 inches. Black and white mohair; glass eyes; open felt mouth; felt pads; all jointed; circa 1950; raised button. $750
Panda Bear: 13 inches. Same description except vinyl pads; 1950s; raised button. $950
Panda Bear: 20 inches. Same description except vinyl pads; 1950s; raised button. Lorraine Oakley collection. $1,400 up

Walking Panda: 7 inches. Plush; plastic eyes; velour pads; made to commemorate President Nixon's visit to China and China's gift of two pandas to Washington, D.C.'s National Zoo in 1972. No I. D. $150 up

Bendy Panda: 3 inches sitting. Black and white mohair; black bead eyes; swivel head; bendable limbs; 1960s; incised button. $200 up

Re-issue Pandas: 10 inches and 15 inches; mohair; open felt mouths; plastic eyes; 1984 only; all I. D. (replica 1938 chest tags). $300-$400

Bengal Tiger: 9.50 inches. Striped mohair; open felt mouth with plastic teeth; green glass eyes; raised button. $350

Pandy: 12 inches. Mohair; glass eyes; felt toes; circa 1955; raised button. $1,000 up

Princeton Bengal Tiger: 15 inches (plus tail). Airbrushed mohair; open felt mouth with four wooden teeth; green and black googly eyes; raised button; 1950s. $450-$500

Lying Tiger: 36 inches. Mohair; green glass eyes; circa 1955; no I. D. $1,000
Musical Tiger: 15 inches. Circa 1955. $525

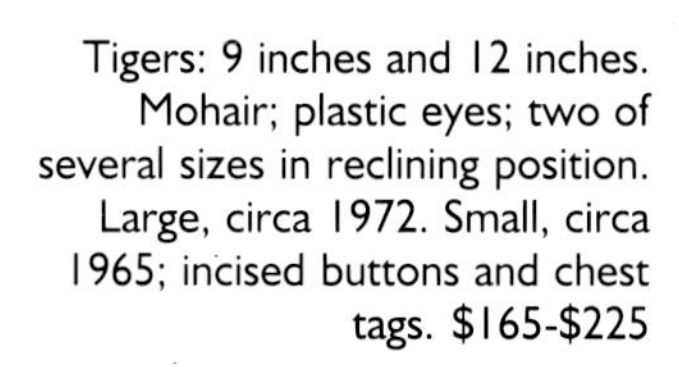

Tigers: 9 inches and 12 inches. Mohair; plastic eyes; two of several sizes in reclining position. Large, circa 1972. Small, circa 1965; incised buttons and chest tags. $165-$225

Tiger Trophy Head: 12 inches. Mohair; glass eyes; open felt mouth with wooden teeth; 1955; raised button. $800 up

Lulac Tiger: 32 inches. Striped mohair; glass eyes; airbrushed feet; long dangling legs; one of a series of "Lulac" animals. Circa 1960. No I. D. $2,200

Floppy Tiger: 9 inches. Mohair; sleep eyes; circa 1950s; all I. D. $150 up

Camel: 6.50 inches. Tan shaded plush; velvet legs; rope tail; black glass eyes; 1950s; chest tag. $125

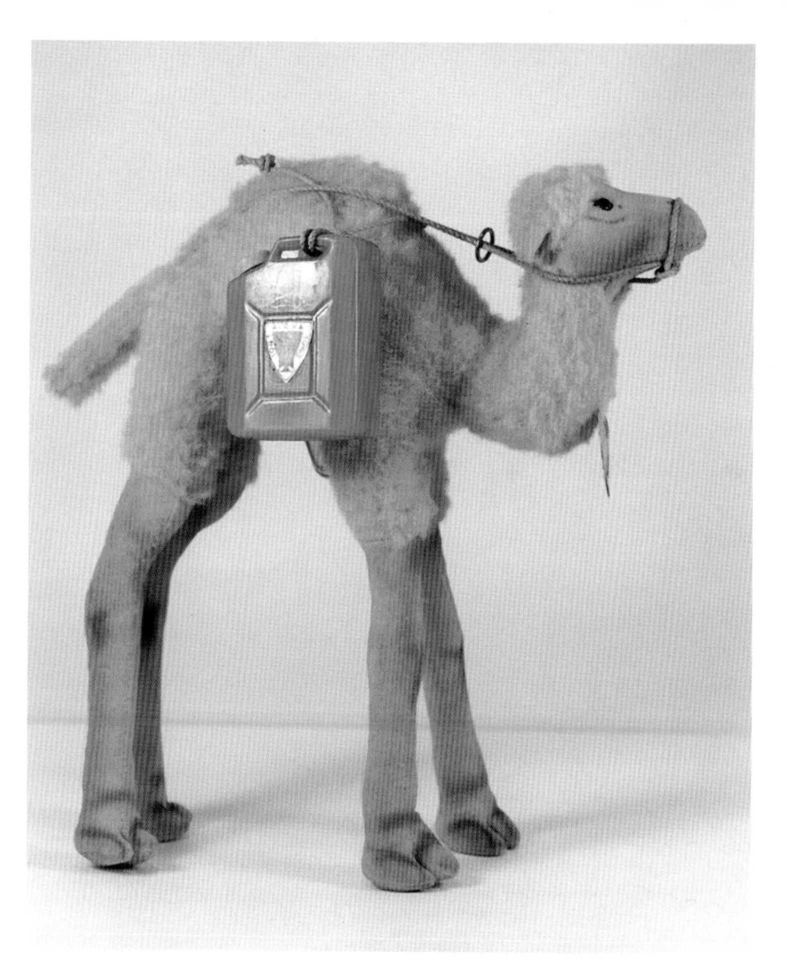

African Camel: 11 inches. Plush; velvet face and legs; glass eyes; rope and plastic water jugs; chest tag; 1940s. $875 up

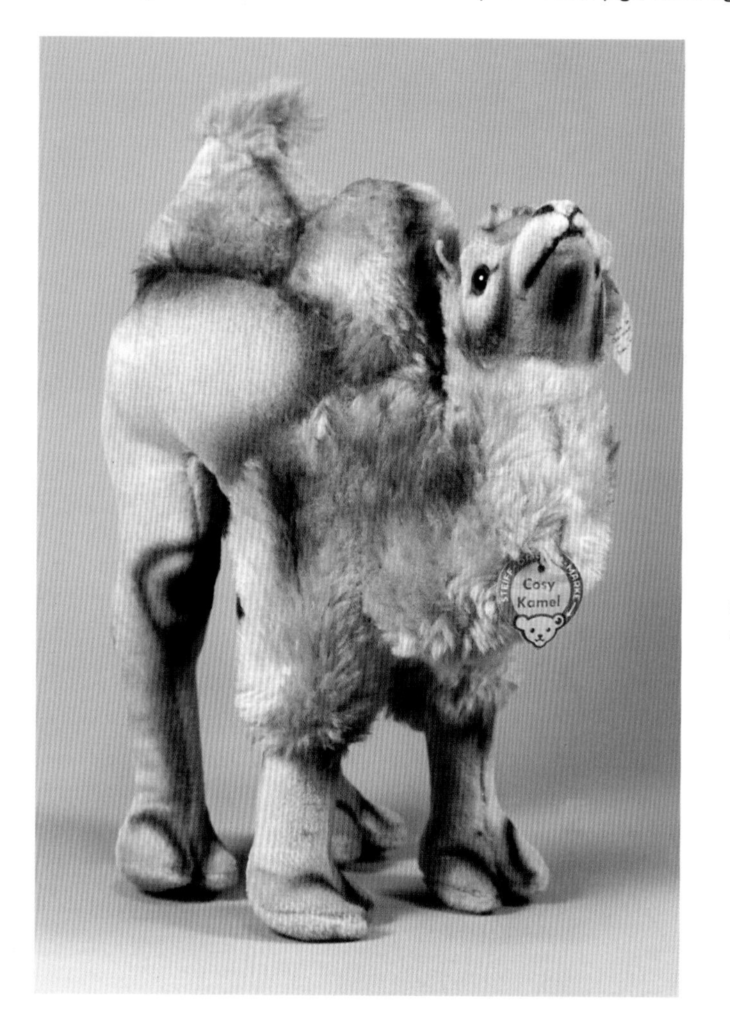

Cosy Kamel: 10 inches. Tan dralon plush; plastic eyes; circa 1960; raised button; chest tag. $175

CHAPTER XV

WATER BABIES

IN THE STEIFF WATERS THERE ARE MORE THAN TWO LITTLE FISHES IN THE BOTTOM OF THE SEA.

When Admiral Byrd's expedition to the South Pole in the 1920s made headlines world wide, the interest in this territory and its wildlife became of major importance. All sorts of characters based on Byrd's exploration soon hit the toy shop scene. It was then that Steiff introduced the penguin. Called *King Peng*, it had the universal mechanism that allowed head movement by levering the tail. Also fitted with a hook in its open mouth, to carry messages, this penguin has a high rarity value since less than 1,700 were made. Other penguins have since been available, notably *Peggy*, a long time favorite.

Many other animals who live in or around the water are also part of the company's product line.

Above: Penguin: 4 inches. White and black mohair; velvet wings; felt beak and feet; glass eyes; somewhat earlier than *Peggy* Penguin; circa 1950. No I. D. $150

Left: Penguin *King Peng*: 12 inches. Brown, white and orange mohair; velvet wings and feet; tri-colored eyes; felt beak; tail moves head in a circular motion; 1920s. Only 1,675 made. Printed FF button; cardboard tag by tail. $3,500

Peggy Penguin: 13 inches. Mohair; felt beak; glass eyes; faux leather feet; swivel head; 1950s. No I. D. $350

Peggy Penguin: 20 inches. White with yellow airbrushing; gray and black mohair; tan mohair feet; yellow and black felt beak; glass eyes; swivel head; 1950s; raised button; chest tag. $450

Wool Ball Penguin: 3.75 inches. Black and white wool; felt beak, feet and wings; glass eyes; swivel head; circa 1960; raised button. $75

Robby Seals: 5 inches and 7 inches. Gray airbrushed mohair; glass eyes; circa 1953; raised buttons; chest tag. $125-$145

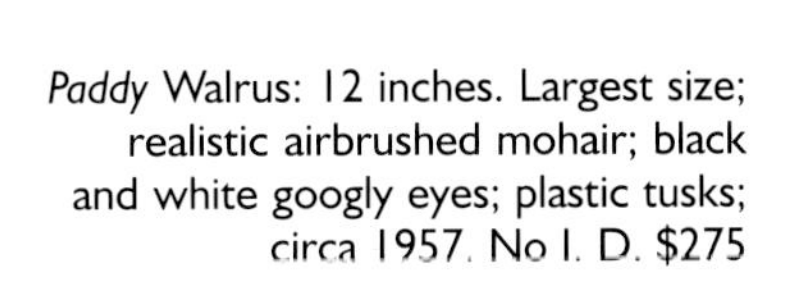

Paddy Walrus: 12 inches. Largest size; realistic airbrushed mohair; black and white googly eyes; plastic tusks; circa 1957. No I. D. $275

Walrus Pajama Bag: 27 inches. Plush; zipper opening; plastic tusks; mohair mustache; glass eyes; 1960s; chest tag. $425

Froggy Frogs: 5, 8 and 12 inches. Larger sizes of airbrushed mohair with glass eyes; smaller one of green velvet with yellow undersides; glass eyes; 1950s and 1960s; raised buttons on larger sizes; incised button on velvet. $110-$150-$225

Froggy: 4 inches. Green and gold velvet. Plastic eyes; circa 1975; incised button; chest tag. $110

Dangling Frog: 13 inches. Velvet; vinyl head; felt lined open mouth; circa 1972; split chest tag. $95
Sitting Frog: velvet; plastic eyes; circa 1965; incised button; bear head chest tag. $125
Lying Frog: 10 inches. Mohair; plastic eyes; circa 1972; incised button; split chest tag. $145

Nelly Snail: 6 inches. Velvet; vinyl antennae and underside; iridescent shell; glass bead eyes; 1950s; raised script button; chest tag. $375 up

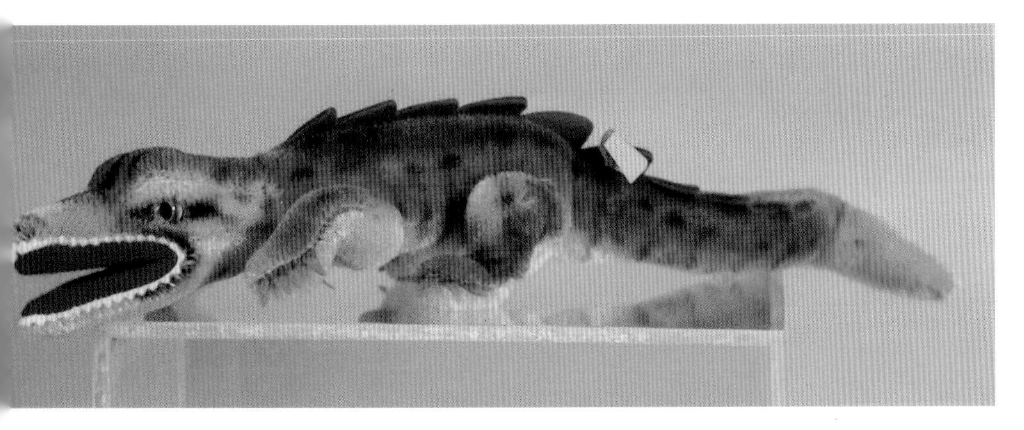

Gaty Alligator: 12 inches. Realistic shaded mohair; open felt mouth and teeth; felt spine; glass eyes. All I. D. $195

Slo Turtle: 13 inches. Colorful mohair; open felt mouth and toes; plastic eyes; circa 1970. $195

Slo Turtle: 5 inches. Mohair; vinyl shell; plastic eyes; circa 1975; incised buttons; chest tags. $85

Nagy Beaver: 7 inches. Short and bristly mohair; open felt mouth, feet and tail; two felt teeth; black plastic eyes; circa 1960s. All I. D. $225

Beaver: 4 inches. Shaded mohair; plastic googly eyes; double weight felt feet; red tie; made for a Canadian airline; circa 1960; raised script button. $225

Cosy Sigi Seahorses: 8 and 11 inches. Plush; glass eyes; 1959; all I. D. $500 each

Swans: 8 inches. White or black acrylic plush; felt beaks and feet; plastic eyes. Production on these examples one year only; 1983; brass button; split chest tag; booklet. $150

Crabby Lobster: 7 inches; felt; glass eyes; pipe cleaner legs; all I. D. 1964. $325

Flossy Fish: 11 inches. Mohair; glass eyes; open felt mouth; 1950s; all I. D. $125

Flossy Fish: 10 inches. White, orange and yellow airbrushed mohair; glass eyes; felt mouth; circa 1960; all I. D. $95-$110

Flossy Fish: 5 inches. Mohair; plastic eyes; felt fins and tail; circa 1975; raised script button; split chest tag. $55

CHAPTER XVI
PUPPETS

STEIFF NOT ONLY HAS THE WORLD ON A STRING, BUT IN THE PALM OF THEIR HANDS AS WELL.

What child doesn't like to play with a puppet? It is a simple toy and yet can stretch imagination to the fullest. Steiff made a marionette bear in 1910 and a plush example fifty years later, but the largest production runs have been the hand puppets. They have ranged from sizes that are so tiny they are meant to be worn on the fingers to ones that fit over the forearm. The finger puppets came out in 1968 and are known as *The Musicians of Bremen.* The four 2.75 inch animals are a donkey, dog, cat and a rooster. Most hand puppets have stayed in production on and off since 1920.

Monkey Puppet: 8 inches. Mohair head and beard; felt face, ears, paws and body; circa 1920. No I. D. $225

Rooster: 10 inches; mohair; felt and plastic trim; 1960s. $140
King Charles: 9 inches. Mohair; 1930s. $145

Clown Puppet: 11 inches. Molded face and hat; felt hands and ears; cotton body; mohair wig; painted mouth, eyebrows, cheeks and nose dab; circa 1930; printed FF button. $400 up

Bully Puppet: 9 inches. Mohair; velvet muzzle; glass eyes; circa 1955; raised button. $115
Molly Puppet: 9 inches. Mohair; glass eyes; circa 1955. No I. D. $90
Mopsy Puppet: 9 inches. Mohair; felt tongue; plastic eyes; circa 1960; chest tag. $120

Above: Snowman Puppet: White wooly plush; felt hat; plastic eyes; felt "carrot" nose; yarn ball buttons; circa 1950. No I. D. $500 up

Left: Lion Cub Puppets: 9 inches. The cub on left is wool plush; circa 1948. $125. The other one is mohair; circa 1955. No. I. D. $95

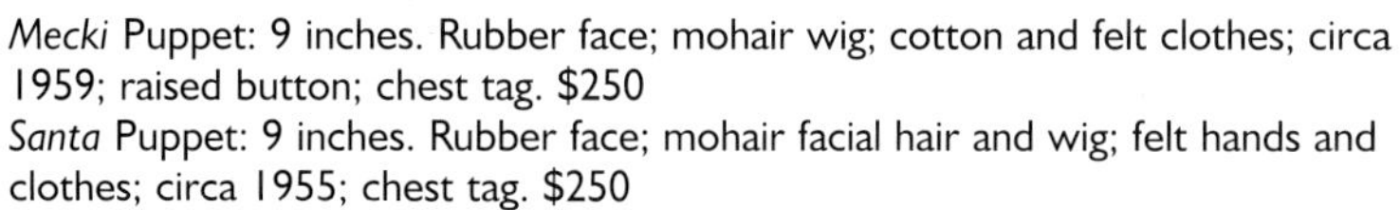

Mecki Puppet: 9 inches. Rubber face; mohair wig; cotton and felt clothes; circa 1959; raised button; chest tag. $250
Santa Puppet: 9 inches. Rubber face; mohair facial hair and wig; felt hands and clothes; circa 1955; chest tag. $250

Fox Puppet: 14 inches. Mohair face, ears and paws; open felt mouth with tongue and two teeth; glass googly eyes; felt tuxedo forms body; supposedly made for F. A. O. Schwarz; circa 1959. Raised button. $295 up

Squirrel Puppet: 9 inches. Mohair; glass eyes; felt paws; 1960s; incised button. $60
Halloween Cat Puppet: 9 inches. Black mohair; plastic eyes; circa 1955; raised script button. $75
Boxer Puppet: 9 inches. Mohair; velvet mouth; plastic eyes; circa 1955; raised script button; chest tag. $125

Finger Puppets: 2.75 inches. "Musicians of Bremen." Raised script buttons; 1960s. $75 each. Raise value if complete set of four.

Gucki Puppet: 9 inches. Rubber face; mohair beard and hair; the balance is felt; circa 1955; raised button; chest tag. $140

Wittie Owl Puppet: Mohair; plastic eyes; 1960s; incised button; chest tag. $135
Lora Puppet: Mohair; plastic eyes; vinyl beak; 1950s; raised button; chest tag. $135

Tiger Puppet: Mohair; plastic eyes, circa 1975; incised button; split chest tag. $95

Baby faced Tiger Puppet: Mohair; glass eyes; 1950s; raised script button. $130

Lion Puppet: Mohair; plastic eyes; 1960s; raised button; chest tag. $120

Smardy Fox Puppet: 9 inches. Mohair; plastic eyes; 1960s; incised button; chest tag. $155

Loopy Wolf Puppet: 9 inches. Mohair; open felt mouth with tongue and teeth; circa 1972; brass button; chest tag. $175

Fox Puppet: 9 inches. Mohair; glass eyes; circa 1955. No I. D. $80

Bear Puppet: 9 inches. Brown and tan mohair; glass eyes; circa 1955. No I. D. $95

Bear Puppet: 11 inches. Plush and velour; plastic eyes; 1970s; brass button. $75

Bear Puppet: 9 inches. Brown and tan mohair; plastic eyes; 1970s; brass button; hang tag; content tag. $125

Froggy Puppet: 11 inches. Velour with plastic eyes and felt hands; circa 1972; brass button. $55
Gaty Puppet: 9 inches. Mohair; open felt mouth; plastic eyes; circa 1955; chest tag. $120
Penguin Puppet: 9 inches. Mohair; felt beak; plastic eyes; 1960s; incised button; chest tag. $105

Wolf Puppet: 11 inches. Plush and velour; 1970s; brass button. $65
Mungo Puppet: 9 inches. Mohair with plush face; circa 1960. No I. D. $65

Loopy Wolf Puppet: 9 inches. Mohair; open mouth with tongue and plastic teeth; 1970s; incised button; split chest tag. $145
Habakuk Puppet: 12 inches. Velour, felt and plush; character made for German market; 1970s; split chest tag. $125

Clown Puppet: Vinyl face; balance felt; 1960s.

Punch Puppet: Vinyl face; felt hat; cotton body; 1950s.

Hansel Puppet: Vinyl face; felt and flannel balance; 1960s; incised buttons; chest tag on *Hansel*. $145 each

Foxy Puppet: 9 inches. Mohair; plastic eyes; 1970s; brass button; chest tag. $105

Snobby Poodle Puppet: 9 inches. Black mohair; tri-colored glass eyes; circa 1960. No I. D. $75

Cockie Puppet: 9 inches. Mohair; glass eyes; circa 1955; chest tag. $115

Robber, Princess and Magician Puppets: 11 inches. Rubber faces; felt bodies and hats; 1970s; chest tag on Robber; incised buttons on Princess and Magician. $125 each

Devil Puppet: 10 inches. Velour, felt and vinyl; incised button; split chest tag; 1970s. $130

Cat Puppet: 8 inches. Black mohair; plastic eyes; circa 1975; incised button; split chest tag. $110

Jocko Puppet: 13 inches. Brown plush; velour face, ears and hands; plastic eyes; 1980s; brass button; split chest tag. $75

Mimic Dog Puppet: 16 inches. Plush; open velour mouth and paws; plastic eyes; formed vinyl nose; arm puppet; 1980s; brass button; split chest tag; hang tag. $135

Monkey Puppet: 12 inches. Plush and velour head; plastic eyes; velour body; 1980s; brass button; split chest tag. $65

Mimic Dally Arm Puppet: 13 inches. Mohair; glass eyes; open mouth with tongue; 1950s. No I. D. $195 up

Happy Puppets: 12 inches. Rabbit, monkey, fox and dog; plush with arms and legs; plastic eyes; ribbon trims; 1980s; brass buttons; split chest tags; hang tags. $100 each

Lamb Arm Puppet: 15 inches. Plush; plastic eyes; 1970s; incised button. $110

Mimic Teddy Arm Puppet: 14 inches. Brown plush; plastic eyes; 1980s; brass button; split chest tag. $165

CHAPTER XVII

DOLLS & HAUTE COUTURE

HOW MAGNIFICENT IT IS TO HAVE, NOT ONLY A PAPER DOLL, BUT A STEIFF DOLL TO CALL YOUR OWN.

One would have to be privy to the Steiff archives to get a look at any dolls prior to the first decade of the twentieth century. A number of dolls in regional costumes were offered in 1892 and it is generally believed that the heads were bought from another firm. In 1897 a variety of dolls, including seven dressed monkeys, were featured in the catalog. Two years later, dolls were vigorously designed and it is interesting to note that they were string jointed in the manner of the first Teddy Bear.

In 1902, the factory began a series of amusing character figures that continued into the 1920s. Large feet, exaggerated knees and a variety of big noses were among the features implemented. The noses were possible to design since the all felt dolls had center face seams. There were military men, national costumed dolls from various countries, a large group depicting cartoon and nursery rhyme personages, clowns of every description, cowboys, schoolteachers, children and a great deal more. The list seems endless.

From 1921 to 1925, Albert Schlopsnies worked with the company producing his own celluloid head design. The faces were softly painted, on the inside, to give a flesh colored glow much like the complexion of the children they were emulating. Since less than 9,200 were manufactured (and presumably the heads were apt to get crushed), the dolls are quite rare.

The 1930s and 1940s saw the emergence of a molded, pressed felt head. These dolls were also modeled after children and bore names such as Ruth, Rosemarie, Hedy and Frieder. Most of these dolls had painted eyes but a small quantity had eyes made of glass.

After the war, and the 1950s especially, a new breed of doll-like figures was introduced. In 1951 Mecki, a dressed hedgehog with a poured latex head, was fashioned after a creation by Ferdinand Diehl. Mecki, his wife Micki and children Mucki and Macki, were heroes of magazine and film fame. Soon Santa Claus and a goodly number of animals incorporated the rubbery heads as well. This period was also the time when an enormous number of charmingly dressed animals took center stage.

Many of these toys remained in the program, because of their popularity, in the following ten years. The next decade saw soft, unjointed dolls and a group of vinyl workmen. Beginning in 1970, beautifully dressed children and storybook characters all strove to keep Steiff in the doll marketplace. During that era, and into next decade, a group of reproduction dolls made it possible for adults to own a sampling of early designs.

Throw Doll: 10 inches. Felt face with center seam; glass eyes; mohair wig; jointed silk arms; velvet body and bonnet; circa 1910. No I. D. $2,000

Above: *Nikolas* Doll: 20 inches. Felt; glass eyes; all jointed, including knees; felt and mohair robe and hat; shown with Struwwelpeter, the book he is featured in (German); 1911; printed FF button; partial white tag. $3,500 up

Right: Dutch Boy Doll: 18 inches. Felt; all jointed; glass eyes; blond wig; felt jacket and trousers; replaced shirt; mend on one clog; circa 1913. No I. D. $1,600 up

Musician Doll: 16 inches. Felt clothes form body; intricately detailed with brass buttons, epaulets and tails; printed leather shoes; mohair hair and mustache; all jointed; shoe button eyes; one of several musicians originally carrying instruments (mandolin is a prop); circa 1913; printed FF button. $3,000 up

ian Doll: 15 inches. Felt; shoe button eyes; mohair wig; jointed; felt and cotton clothes; missing moccasins; 1; printed FF button. $1,750 up

Tyrolean Doll: 14 inches. Felt; glass eyes; circa 1911; printed FF button. $1,200 up

Inge Dutch Doll: 11 inches. Felt; mohair wig; blue glass eyes; dressed in national costume of Holland; all original except for replaced cap; 1909; printed FF button. $1,200 up

Tyrolean Boy Doll: 13 inches. Felt; glass eyes; cotton and wool clothes; leather shoes; circa 1914; printed FF button. $1,900

Cowboy Car Ornament: 7 inches. Felt and rayon; glued-on nose and mouth; rope lariat; circa 1948; raised script button; U S Zone tag. $900

American Soldier: 14 inches. Felt; braid trim; leather belt and shoes; circa 1909; printed FF button. $2,000

Santa Claus: 12 inches (on right). Vinyl face; felt hands and clothes; mohair beard and plush trim; 1950s (re-issue on left described elsewhere). Chest tag; raised button on plastic bracelet. $500

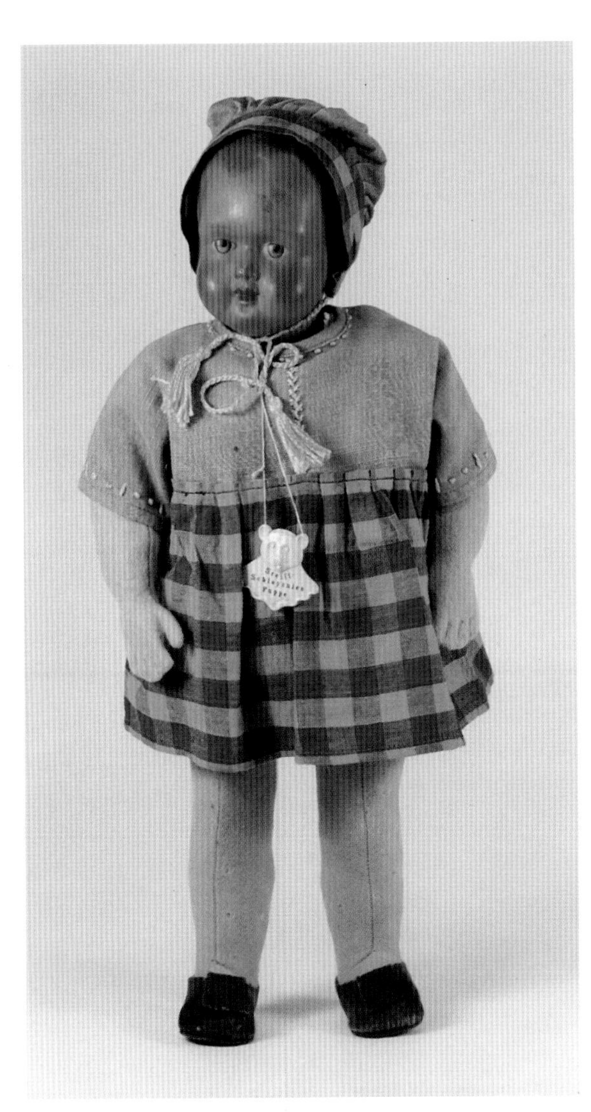

Vera Schlopsnies Doll: 15 inches. Named after the designer; celluloid head painted from the inside and then stuffed with excelsior; felt body; jointed; original cotton clothes and leather shoes; to have hang tag intact is rare; 1921-1925 only. $4,000 up

Lisl: 14 inches. Molded felt face; glass eyes; mohair wig; fabric arms; felt body and legs; felt and cotton clothes; 1938; chest tag. $1,500 up

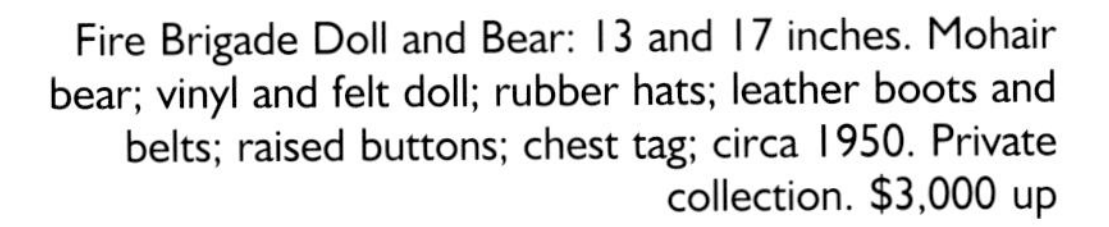

Fire Brigade Doll and Bear: 13 and 17 inches. Mohair bear; vinyl and felt doll; rubber hats; leather boots and belts; raised buttons; chest tag; circa 1950. Private collection. $3,000 up

Duke University Mascot Devil: 11 inches. Rubber face; felt forms body and clothes; replaced cloak; raised button; U S Zone tag; circa 1950. $700 up

Doll *Mat*: 8 inches. Rubber face; felt body; cotton and felt clothes; paper and wood accordion; circa 1950; raised button; chest tag. $500

Clownie: 5 inches. Rubber head and arms; felt and cotton clothes; some melting on hands; circa 1950; raised button; chest tag. $175

Clownie Doll: 7 inches. Rubber painted face; mohair hair; felt jointed body; non-removable cotton clothes; felt hat and gloves; circa 1955; raised button on plastic bracelet; chest tag. $175

Neander Cave Men: 5 inches and 8 inches. Small example is vinyl; larger has vinyl head and all jointed felt body; long mohair hair; mohair sarongs; felt shoes; plastic tooth on chain around neck; circa 1959; raised buttons/stock tags; bear head chest tags. $275 (small); $325 (large)

Gucki and *Lucki* Dwarves: 11 inches. Rubber faces; felt bodies; cotton clothes; vinyl shoes; felt cap has feather attached with a ribbon; circa 1958. Two of three; raised button attached to vinyl wrist bracelet; chest tag. $215

Gucki, Pucki and *Lucki Dwarves:* 7 inches. Smaller version of the 11 inch dwarves. All I. D. 1950s. $150 each

Pucki, Lucki and *Gucki* Dwarves: 4 inches. Rubber faces and hands; felt clothes; raised buttons; chest tags; 1950s. $195 set of three

Dwarf: 4 inches. Rubber face and hands; felt clothes; advertising item for Echt Stonsdorfer Co. Circa 1960; raised button; chest tag. $295

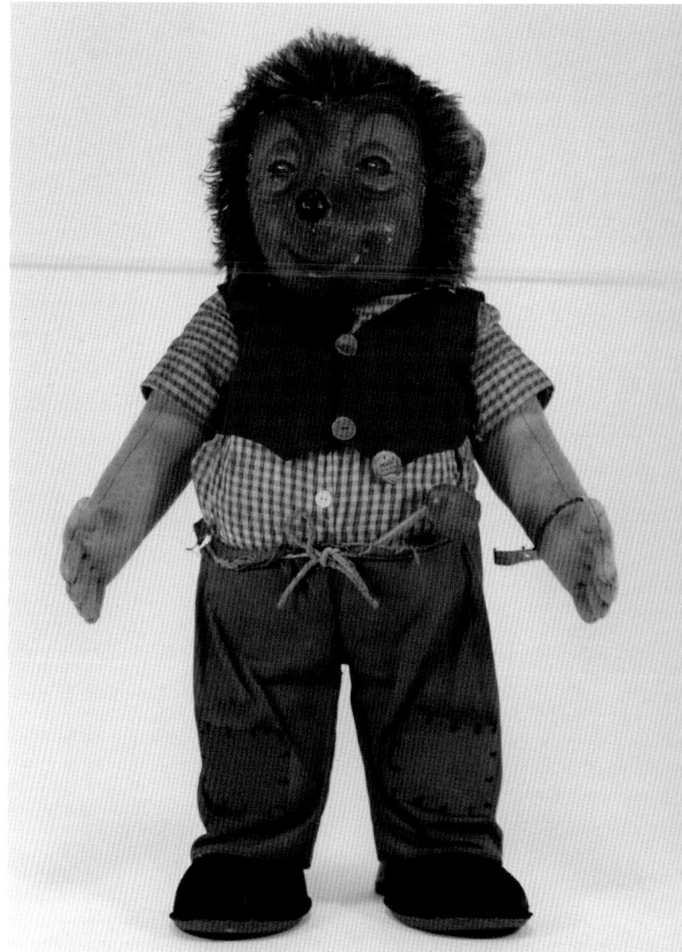

Mecki Hedgehog: 20 inches. Rubber face (early edition, therefore crackled); felt body, vest and shoes; cotton shirt and pants; mohair wig; all jointed; circa 1950. Diehl film character; raised button on wrist bracelet; chest tag. $150

Sandy Doll: 11 inches. Vinyl head; painted eyes; felt body and made-on clothes; carries a cotton bag full of "star dust;" circa 1960; incised button; chest tag. $225

Tele Maxl Doll: 12 inches. Rubber head; painted features; mohair wig; felt body; felt and cotton clothes; all jointed; made especially for a television company in Munich; 1958; raised script button on bracelet. $375 up

Uwe Doll: 15 inches. Velour and cotton; plastic eyes; 1970s; incised button; split chest tag. $165

Mucki and *Macki*: 4.50 inches. Vinyl with cotton and felt clothes; children of the hedgehog family so popular in Germany; Diehl film characters; 1970s; raised button; split chest tag. $55 each

Workmen and Sport Dolls: 7 inches. Vinyl with felt, cotton and velour non-removable clothes and accessories; 1970s; raised buttons; split chest tags with doll's names. $185 each

Re-issue Fireman Doll: 15 inches. Vinyl head and helmet; felt uniforms; great attention to detail of equipment (ropes and hatchet hang from belt in back). Made as a promotion for a German fire company in the 1980s; brass button. $250

Little Red Riding Hood and Wolf: 9 inch doll and 5 inch wolf. Limited to 2,000 pieces in 1987. In box with chest tag on wolf; Suzanne Gibson tag on doll. $295

Santa Claus: 11 inches. Rubber face; felt clothes, hands and bottom of soles; plush beard and suit rim; 1984 only; brass button; stock tag; chest tag. $295
Santa Claus: 7.50 inches. Same description as larger size except vinyl shoe soles; both are replicas of 1953 issues. Brass button; stock tag; chest tag; hang tag. $225

Gusto Clown: 13 inches. All felt; black glass eyes; long plush hair; painted clown face; finely detailed non-removable clothes; all jointed; reproduced from a figure in a 1910 circus display for a Berlin Department store (created by Albert Schlopsnies). Limited edition of 1,000; 1994; brass button; white stock tag on coat; in presentation box with certificate. $450

Uncle Sam Doll: 19 inches. Velour, felt and plush; dressed in patriotic colors; glass eyes; arms and legs are jointed; reproduced from a doll registered in 1904 and modeled after a cartoon seen in the U. S. newspaper the *Daily Dispatch.* Limited edition of 1,000; 1994. Tiny button in left ear; larger brass button and white stock tag on coat tail; five buttons on front; in presentation box with certificate. $400

Coloro Clown: 17 inches. Felt; replica of a 1911 figure. 1988-1990. $400

Noso Clown: 17 inches. Felt. Replica of a circa 1911 figure. 1995. $360

Musical Teddy Baby: 9 inches. Brown mohair; tan inset mohair muzzle; open mouth; cylinder body that activates music box when pushed down; original dress; circa 1950; raised button; U S Zone tag. Photo courtesy of Christie's, London. Sold at Christie's auction Dec. 1996.

Musical Cat: 9 inches. Mohair; tubular body; press down to activate music box; glass eyes; original dress; circa 1950; No I. D. $695

Manni Rabbit: 4 inches. Beautifully dressed in old laces, trims and rhinestones. Tag under skirt reads "Helen Ratkai is my couturier." No I. D. $250

Snobby Poodle: 5 inches, White mohair; brown glass eyes; black glass nose; felt tongue; all jointed; dressed for Christmas in velvet; pearl earrings and crown wreath; tag on skirt reads "Helen Ratkai is my couturier." Circa 1950. No I. D. $250

Bazi Dogs: 9 inches. Pair of dressed Dachshunds; boy as hunter; girl Tyrolean; mohair and cotton; felt and cotton clothes; glass eyes; circa 1950. No I. D. $1,200 pair

Pippy Mouse: 4 inches. Velvet head; felt ears; bead eyes; rubber arms, legs and tail; non-removable clothes; circa 1950; chest tag. $450

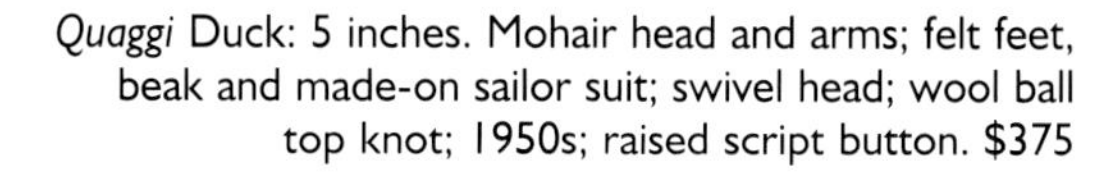

Quaggi Duck: 5 inches. Mohair head and arms; felt feet, beak and made-on sailor suit; swivel head; wool ball top knot; 1950s; raised script button. $375

Nikili Rabbits: 8 inches. Mohair; glass eyes; swivel heads and arms; open felt mouths; felt and cotton clothes; circa 1959; raised script buttons. Bear head chest tags. $900 pair

Rico Ski Bunny: 17 inches. Tan mohair with red plush; plastic eyes; all jointed; felt pads; blue scarf; ski poles; incised button; chest tag. 1960s. $300

Hupfi Rabbit: 6 inches plus ears. Ski bunny with mohair head; paws, felt; made-on clothes; plastic eyes; 2 poles; 2 skis; incised button; chest tag. $325
Knupfi Rabbit: 6 inches. Ski bunny injured; made like *Hupfi* except arm in sling; cast on leg; one pole; one ski; incised button; chest tag. $325

Pieps Mouse: 3.50 inches. White mohair mouse with pink eyes and black nose; felt ears, paws and tail; dressed in beribboned net, flowers and a rhinestone necklace by F. A. O. Schwarz. 1960s; raised button. $200

Pieps Mouse: 3.50 inches. White mohair; pink eyes; bead nose; felt paws; dressed as a clown by F. A. O. Schwarz; circa 1960; raised script button. $300

Pieps Mouse: 3.50 inches. White mohair; pink eyes; bead nose; felt paws; dressed as a bride by F. A. O. Schwarz; circa 1960; raised script button. $300

Pieps Mouse: 3.50 inches. White mohair; pink eyes; dressed in Spanish costume and sold by F. A. O. Schwarz; circa 1960; raised button. $300

Ballerina Pieps Mouse: 3.50 inches. White mohair; pink eyes; black bead nose; double weight felt pads; dressed in satin, lace, net, rhinestone necklace and flower headpiece by F. A. O. Schwarz; one in a series of dressed mice; circa 1960; script button. $250

Dressed Dalmatian: 5.50 inches. Black on white spotted mohair; black glass eyes; the only Dalmatian in begging position; wears red taffeta cape lined in white and tied with gold cord; gilt crown with red and white "jewels," made for F. A. O. Schwarz; 1960s; chest tag; F. A. O. Schwarz tag. $950

Advertising Rat: 15 inches. Gray flannel; zippered racing suit; made for Lauda-Air; 1997; brass button; hang tag. $190

CHAPTER XVIII
WHEELED TOYS
MERRILY THEY ROLL ALONG.

Steiff toys on wheels happen to be among my favorites. Some are large enough to hold a Teddy Bear rider and what could be better than that? The first wheels were made of cast iron. They were usually left in the natural state, but were occasionally painted bronze. Wooden wheels came next in a varnished finish, but the somewhat later eccentric models had painted ones. Eccentric movement resulted from the wheels being offset, causing the toy to go up and down when pulled. By the 1930s, wheels could be rubber tired metal, but the preponderance of these playthings are most likely to be from an era twenty years later.

Bear on wheels: 22 inches. Off-white mohair; shoe button eyes; swivel head; cast wheels; circa 1908; blank button. $2,500 up

Bear on wheels: 24 x 30 inches. Brown mohair; clipped tan mohair inset muzzle; glass eyes; rubber and metal wheels; original pull cord; pull growler; 1949. U S Zone tag. $1,850

Cat on cast wheels: 9 inches. Felt; glass eyes; replaced collar; circa 1910; no I. D. $1,300 up

Cat on cast wheels: 12 inches. Long white mohair; velvet inner ears; glass eyes with rims; circa 1910; printed FF button. $1,600 up

Cat on painted bronze cast wheels: 9.50 inches. Mohair; glass eyes; original bell; ribbon imprinted with Steiff; circa 1910; printed FF button; white stock tag. $1,600 up

Cat on eccentric wheels: 9 inches. Faded mohair; glass eyes; circa 1925; printed FF button. $895

Tige on wheels: 17 inches. Brown coat wool; shoe button eyes; shows some wear; circa 1910. No I. D. $600 up

Bulldog on eccentric wheels: 9 inches. Airbrushed felt; embroidered teeth and nose; shoe button eyes; circa 1913. No I. D. $1,000 up

St. Bernard on cast wheels: 16 inches. White and brown mohair; glass eyes; pull string bark (still operating); circa 1910; printed FF button. $1,400

Waldi on eccentric wheels: 13 inches. Long rusty gold mohair; short mohair snout and legs; glass eyes; embroidered nose, mouth and claws; original collar; hops when pulled; some wear on muzzle; circa 1930; printed FF button. $350

Dachshund on eccentric wooden wheels: 13 inches. Rust felt; glass eyes; hopping action when pulled; circa 1920; printed FF button. $950

Foxy Dog on wheels: 7 inches. White mohair with airbrush markings; glass eyes; red wooden wheels; circa 1950. $450

Molly on eccentric wheels: 9 inches. Mohair; glass eyes; swivel head; red wooden wheels; circa 1930; printed FF button. $700

Fox Terrier on eccentric wheels: 9 inches. Silky plush; glass eyes; circa 1935; printed FF button; square head bear tag. $595

Bully on eccentric wheels: 5 inches. Airbrushed mohair; velvet muzzle; glass eyes; collar; wheels offset to produce hopping action; all I. D. 1950s. $395

Hoppel-Dachel: 12 inches. Mohair; open mouth with felt tongue; glass eyes; stuffed loose in the middle so dog wiggles when pulled; removable (via Velcro strips) from rubber tired cart; circa 1960; raised button; chest tag. $700

Record *Hansi*: 10 inches. Blonde mohair; glass eyes; felt pads; fully jointed; on bellows cart that presents pumping action when pulled; 1948; raised script button; U S Zone tag. $750

Rooster on eccentric wheels: 10 inches. Colorful felt; glass eyes; wheels offset to provide a hopping action when pulled; circa 1910; tiny printed FF button placed on the wattle. $1,400 up

Duck on cast wheels: 11 inches. Velvet with airbrushing; felt feet; shoe button eyes; 1908. No I. D. $1,300 up

Duck on cast wheels: 11 inches. Colorful felt; velvet head; glass eyes backed by felt; circa 1910; *Steiff* printed on neck ribbon. $1,200 up

Duck on eccentric wheels: 11 inches. Felt; velvet head; yarn pom pom; shoe button eyes backed by felt; "Steiff" imprinted on wheels; circa 1914. $1,000

Duck on wheels: 4 inches. Airbrushed mohair; orange felt beak and feet; glass eyes; circa 1950; raised script button. $275

Ducks: 3.50 and 5.50 inches. Gray and green felt; black bead eyes; gold embroidered trim; mother duck is followed by 5 babies; circa 1920. No I. D. $1,000

Goose on eccentric wheels: 10 inches. White mohair; felt beak and feet; glass eyes; 1950s. No I. D. $500

Donkey on cast wheels: 30 inches. Gray mohair; glass eyes; original leather saddle and trappings; wheels turn in the front; shows wear; circa 1920. No I. D. $1,500 up

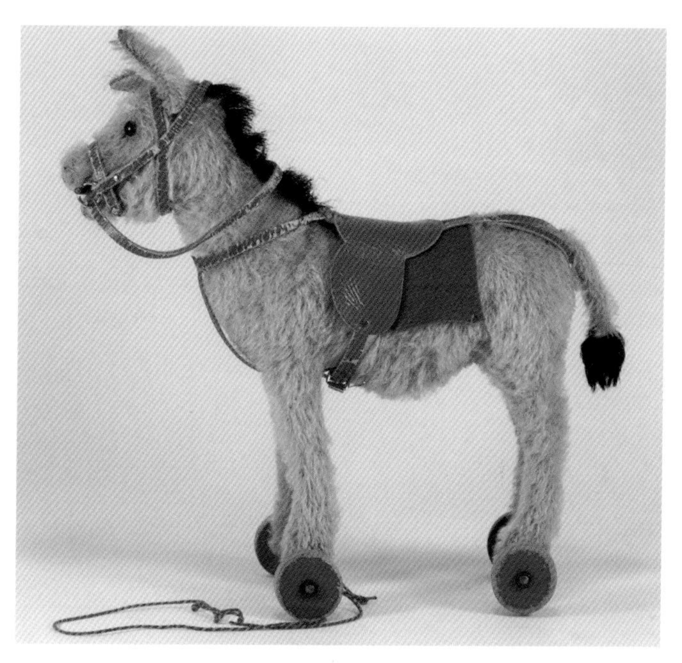

Donkey on wooden wheels: 13 inches. Gray mohair; horsehair mane and tail pom pom; shoe button eyes; felt blanket; leather saddle and bridle; original pull string; circa 1920. No I. D. $800

Swan on cast wheels. 13 inches. Black velvet head and neck; felt beak, feet and body with wings pieced to simulate feathers; glass eyes; silk ribbon; circa 1910. No I. D. $2,000

Donkey on wooden wheels: 10 inches. Gray, white and black plush; plastic eyes; felt saddle; 1980s; brass button; split chest tag. $125

Horse on cast wheels: 19 inches. Felt; shoe button eyes; leather working horse halter. No I. D. Circa 1915. $1,300

Horse on wooden wheels: 15 inches. Felt; glass eyes; felt blanket; horsehair mane and tail; small repair by left eye; leather saddle; circa 1920; printed FF button. $1,500

Horse on wooden wheels: 14 inches. Orange and white mohair; horsehair mane and tail; leather hooves; open felt mouth; velvet blanket; show-horse stance; circa 1925; FF button; full orange stock tag. $1,200

Horse on rubber tired metal wheels: 24 inches. Airbrushed mohair; glass eyes; leather hooves, saddle and trappings; felt blanket; circa 1950; raised button. $1,200

Goat on cast wheels: 24 inches. Off-white curly mohair; green glass eyes; leather horns; circa 1910; printed FF button. $1,900 up

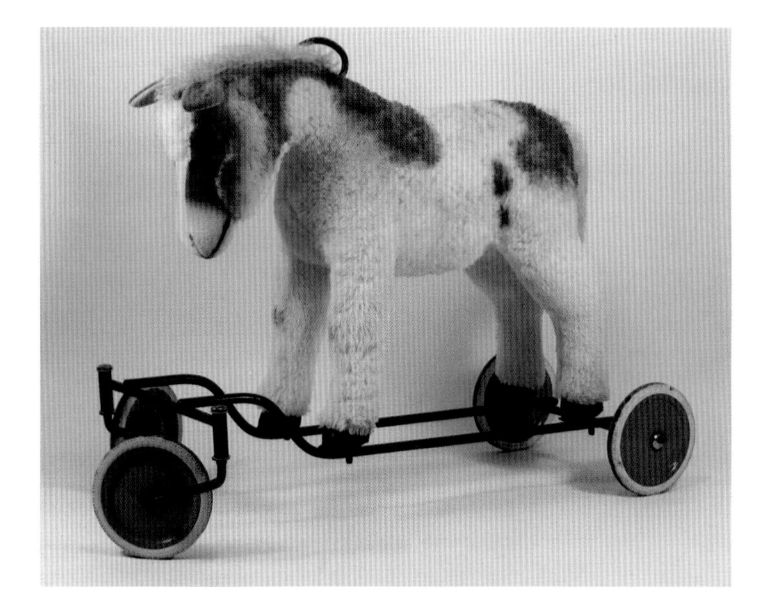

Sheddy Horse on wheels: 20 inches. Plush; plastic eyes; on metal frame with rubber tired wheels; steering handle; 1860s; raised button. $450

Horse on wheels: 26 inches. Brown and white plush; glass eyes; original trappings and saddle; removable rockers; 1960s; Steiff 100 on the rubber tires. $1,100

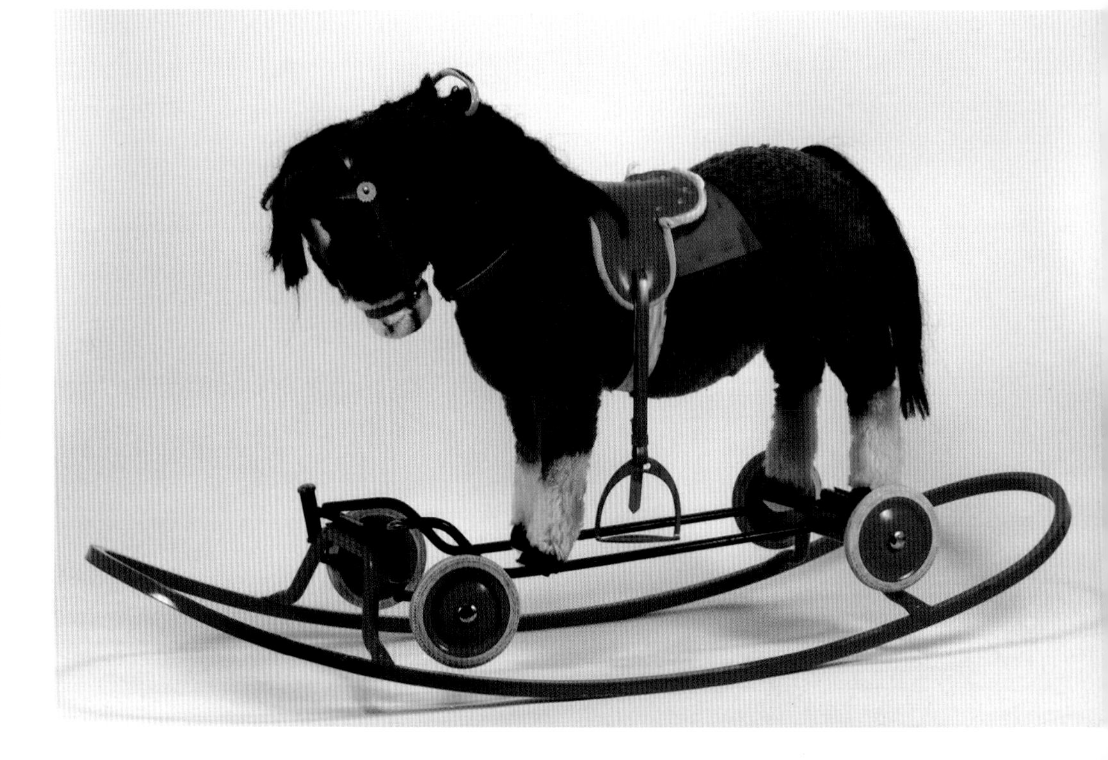

Goat on eccentric wheels: 12 inches. Off-white mohair; glass eyes; goat's head moves up and down via attached metal piping that runs from cart to neck; unusual; circa 1929. No I. D. $800 up

Zicky Goat on eccentric wheels: 7 inches. Airbrushed mohair; green glass eyes; rubber horns; original ribbon and bell; 1950s; raised script button; chest tag. $395

Lamb on wooden wheels: 12 inches. Curly wool; shoe button eyes; pull string and ribbon. No I. D. 1920s. $850

Lamb on wooden wheels: 15 inches. Wooly plush; blue glass eyes; leather hooves; original ribbon and bell; circa 1930; printed FF button. $1,150

Cow on cast wheels: 11 inches. White felt with airbrushed markings; shoe button eyes; leather collar; circa 1910; Printed FF button with full white stock tag. $1,600

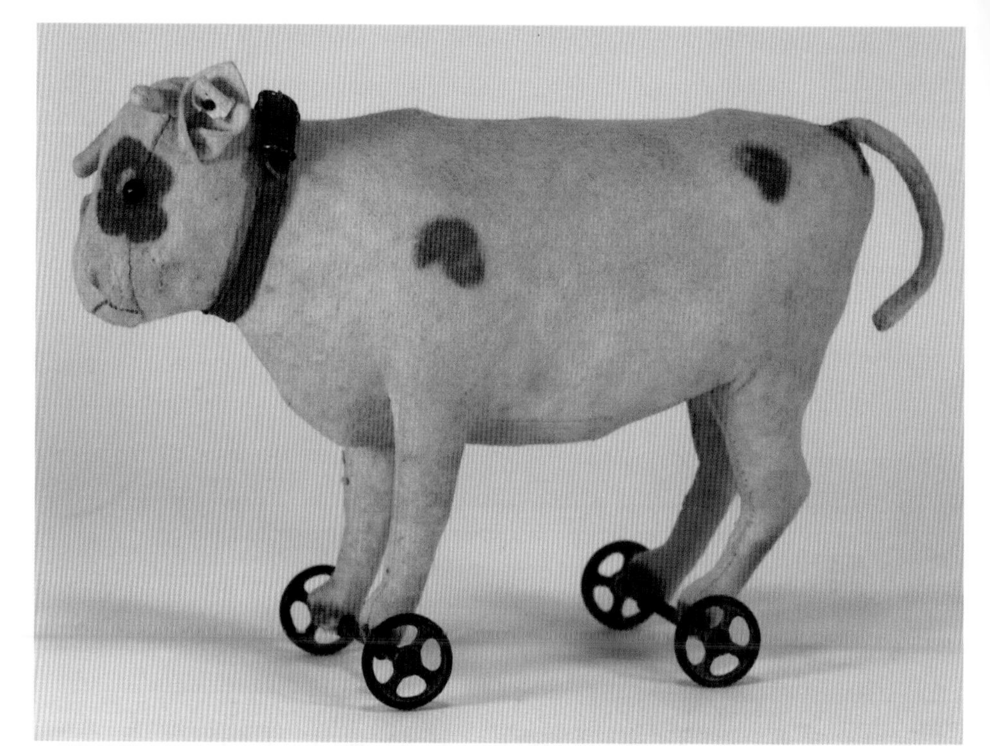

Cow: 20 inches. Mohair; shoe button eyes backed with white felt; leather horns and hooves; working pull "moo;" leather collar and bell; front axle turns; circa 1930; printed FF button; trace of white stock tag. $1,800 up

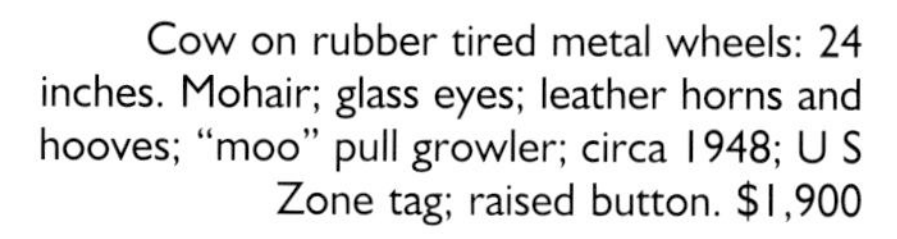

Cow on rubber tired metal wheels: 24 inches. Mohair; glass eyes; leather horns and hooves; "moo" pull growler; circa 1948; U S Zone tag; raised button. $1,900

Deer on cast wheels: 14 inches. Napped wool; felt horns and inner ears; shoe button eyes; blanket not original; circa 1910; FF button. $850

Buffalo on cast wheels: 26 inches. Cinnamon brown and long gray mohair; felt horns and feet. Shoe button eyes; wheels turn in the front; wear to mohair; circa 1910. No I. D. $2,000

Bison on cast wheels: 18 inches. Brown felt and mohair; glass eyes; restored horns and tail; wheels painted bronze; circa 1910. No I. D. $1,500 up

Bison on cast wheels: 24 inches. Brown felt; mohair mantle and front legs; shoe button eyes backed by felt; wheels painted bronze; shows wear; circa 1910. No I. D. $1,500 up

Camel on cast wheels: 9 inches. Brown felt with mohair accents; shoe button eyes; felt blanket; leather bridle; bronze painted wheels; moth holes in felt; circa 1910; printed FF button; trace of white stock tag. $900

Camel on cast wheels: 15 inches. Brown felt; gray mohair on humps, head, neck and upper front legs; shoe button eyes; tail restored; felt blanket replaced; shows wear; circa 1912. No I. D. $800

Camel on cast wheels; 18 inches. Mohair; shoe button eyes; original blanket; circa 1912. No I. D. $2,000

Monkey on eccentric wheels: 8 inches. Brown and white felt; shoe button eyes backed by felt; felt hat with pom pom; felt ruff with bell; varnished wheels that are offset to produce hopping action when pulled; circa 1912. No I. D. $700

Monkey on eccentric wheels; 9 inches. Felt; shoe button eyes; circa 1920; Steiff imprinted on all four wheels. $1,200

Coco on eccentric wheels; 9 inches. Gray mohair; long white mohair around face; felt face and ears; green glass eyes; red felt rear end and hat; swivel head; hopping action when pulled; 1950s; raised button. $425

Record Peter: 9 inches. Brown mohair; felt inset face, ears and paws; glass eyes set in lids; bellows operating; when pulled pumping action occurs and bellows squeak; circa 1948; raised button; cloth U S Zone tag; chest tag. $500

Elephant on cast wheels: 15 inches. Gray mohair; shoe button eyes backed by felt; felt tusks; original felt blanket and trappings; leather saddle; circa 1910; printed FF button. $2,500

Elephant on cast wheels; 8 inches. Gray felt; shoe button eyes backed by felt; felt tusks and original red felt saddle; circa 1910. No I. D. $1,300

Elephant on cast wheels: 10 inches. Gray felt; shoe button eyes; white padded felt tusks; wheels painted bronze; original red felt blanket; circa 1910; printed FF button; trace of white stock tag. $1,300

Elephant on wooden wheels: 13 inches. Charcoal gray felt; glass eyes; rayon covered tusks; replaced felt blanket; pull growler; circa 1920; printed FF button. $700

Elephant on wooden wheels: 7 inches. Felt; white felt tusks and eyes backing; black glass eyes; replaced blanket; 1920s. $450

Elephant on rubber tired metal wheels; 20 inches. Gray mohair; black glass eyes; felt tusks; shows wear; circa 1950; head decoration not original; raised button. $500

Elephant on wooden wheels: 30 inches. Gray felt; shoe button eyes backed by pink felt; felt tusks; original felt saddle and trappings; circa 1920. No I. D. $2,000

Elephant on rubber tired metal wheels: 22 inches. Mohair; glass eyes backed by felt; felt tusks; leather head trappings; felt blanket; circa 1950; raised button; U S Zone tag. $1,400
Poodle on rubber tired metal wheels; 21 inches. Mohair; humanized glass eyes; leather collar; pull growler; circa 1950. $1,400

Above: Lioness on wooden wheels: 14 inches. Airbrushed mohair; glass eyes; circa 1925; printed FF button. $600

Right: Giraffe on cast wheels; 12 inches. Airbrushed felt; shoe button eyes; circa 1910. No I. D. $900

Tiger on rubber tired metal wheels: 24 inches. Mohair; glass eyes; open felt mouth; wooden teeth; 1950s; raised button. $1,400

Wooden bird on eccentric wheels: 5.50 inches. Painted bird with glass eyes; offset wheels causes jumping action when pulled; circa 1920; printed FF button; trace of white stock tag imbedded near eye. $750

Riding Lady Bug: 20 inches. Realistic colored mohair; glass eyes; rubber tired metal wheels; front axle turns; holding bar; pull string; 1960s; raised button. $500

Wood Dachshund and Lion on wheels: 8.50 inches. Cut out shape with wood burned design; red painted wheels; circa 1970; Steiff brass logo on the sides. $150 each

Wood Bison and Camel on wheels; 9 inches. Wood in silhouette shape; wood burned design; circa 1970; Steiff brass logo on sides. $150 each

Wooden Elephant and Rabbit on wheels: 7.50 and 10 inches. Cut out shape with wood burned design; red painted wheels; circa 1970; Steiff brass logo on sides. $150 each

CHAPTER XIX
NOVELTIES

IT'S NOT UNUSUAL TO FIND THE UNUSUAL AMONG STEIFF'S PLAYTHINGS.

Although the Steiff Company is known primarily for its soft toys, they have, over the years, presented playthings in other mediums as well. Blocks, kites, wagons, trucks and tractors are just a few of the many things encountered bearing a Steiff logo.

It is certainly interesting to see prototypes that were never manufactured, even though one can't ordinarily purchase them. Aside from this type of novelty, there are other oddities that are available if one searches hard enough. Certainly *Peck* the germ falls into this category.

Hide a Gifts (designed to do exactly what the name implies), footstools, skittles and a host of other wonders help to make the name Steiff synonymous with imagination.

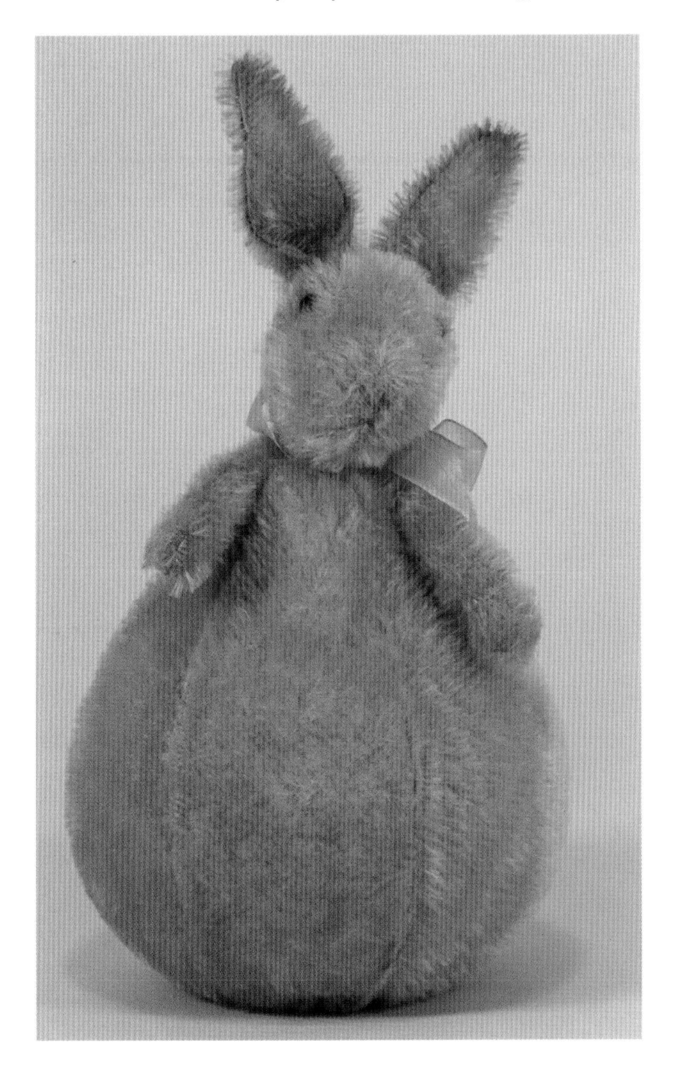

Above: Marionettes: Three Steiff bears modified to be used as marionettes by a puppeteer in the 1930s. Photo courtesy of Christie's, London.

Left: Roly Poly Rabbit: 8.50 inches (plus ears); pale pink and white mohair; swivel head and arms; shoe button eyes backed by felt; weighted body with inner chimes; printed FF button; 1920s. $3,000 up

Skittle Set: Bear "King pin" dressed in felt; elephant, rabbit, cat, pig, poodle, bulldog and two other dogs are all made of velvet; coat wool type fabric on bear and curly material on the poodle; circa 1895; (pre-button). No I. D. $25,000

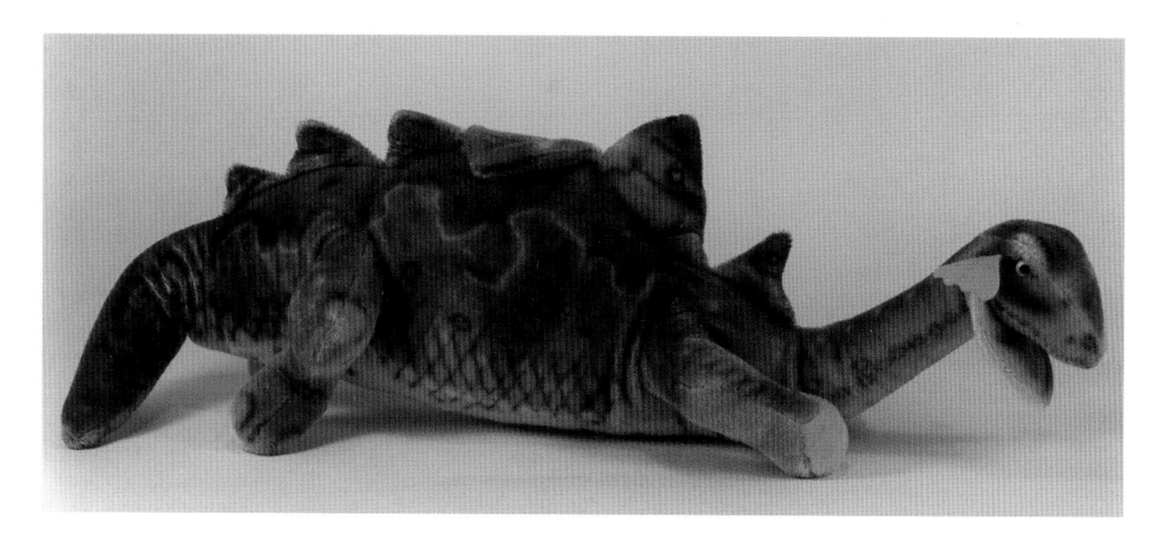

Dinos Dinosaur: 25 inches. Airbrushed mohair; open felt mouth; felt ears; glass eyes; 1950s; raised script button. $1,500

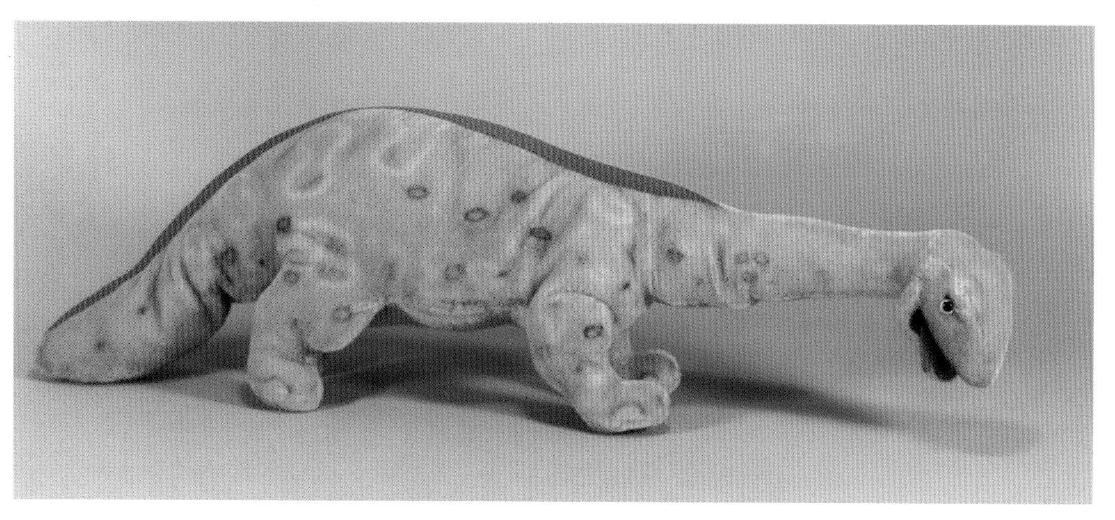

Brosus Dinosaur: 30 inches. Airbrushed tan mohair; felt spine and tongue; glass eyes; open mouth; circa 1955. No I. D. $1,000-$1,100

Tysus Dinosaur: 17 inches. Colorful airbrushed mohair; felt open mouth; glass eyes; swivel arms; felt spine; circa 1955; raised button. $1,000 up

Starfish Footstool: 19 inches. Colorful airbrushed mohair; metal frame and legs; some repairs; very hard to find; circa 1950; U S Zone tag. $850 up

Peck Germ: 3 inches. Green mohair head; felt covered wire limbs; made for German pharmaceutical company; no I. D. $1,200 up

Turtle Footstool: 19 inches. Mohair; on metal frame with metal legs; circa 1950; raised button. $600

Frog Footstool: 22 inches. Colorful and realistic airbrushed mohair; large glass eyes; metal armature; circa 1955. No I. D. $600

Ball on rope, Hammer and Drumstick: Mohair; mohair rope on ball that has inner bells; circa 1955; raised buttons on all three; chest tag on drumstick. $145 each

Hide-a-Gifts: 6 inches. Mohair heads and arms (except plush fox); swivel heads; plastic eyes; felt dresses hollow underneath; novelty to hide small gifts; 1960s; all I. D. $145 up each

Hide-a-Gift Bear: 6 inches. Same description as the animals. 1960s. All I. D. $160

Baby Toys: 4, 6 and 6 inches. Red, yellow and blue ring; colorful block with various Steiff toys lithographed on all sides; velour butterfly; 1950s, 1970s and 1980s. Buttons and tags of the periods. $45-$75

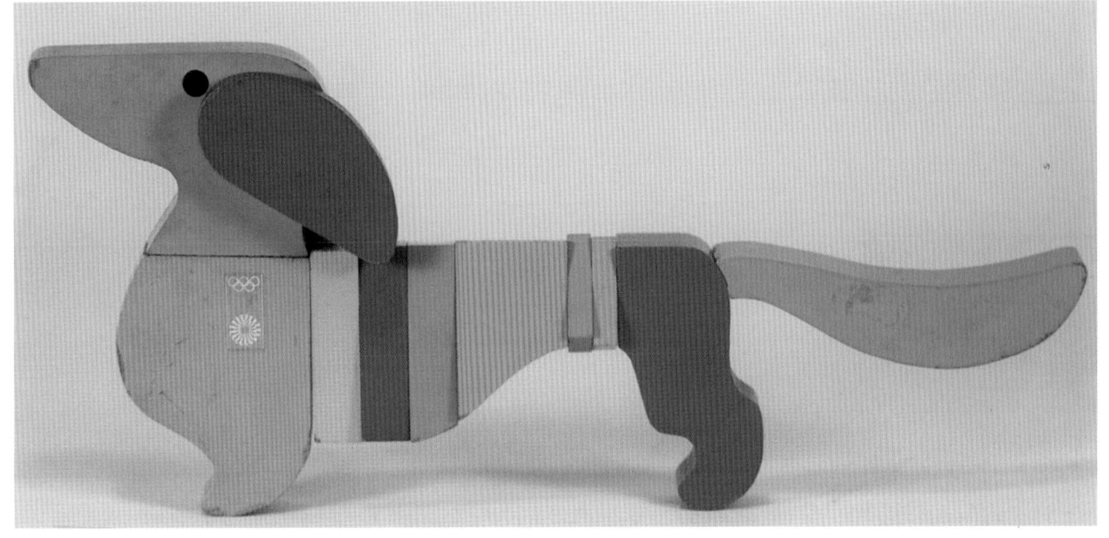

Olympic *Waldi*: 13 inches. Colorful dog in sections like a puzzle; made for the 1972 Olympics in Munich. Steiff logo on one side; Olympic insignia on the reverse. $150

Above: Olympic Pocketbook Prototypes: 7 inches in diameter. Plush with zipper closure. Never in production. The only examples ever made.

Left: Prototype Raccoon: 11 inches. In 1978 and 1979 Steiff, through their U. S. representative, made presentations to the American Olympic Committee in order to obtain a license to make 1980 Olympic souvenirs. Three prototypes were made incorporating the Lake Placid mascot and Olympic logo. Although the samples were attractive, an official license was never granted. These are the only examples ever made.

Jeany Backpack: 17 inches. Plush bear in form of a denim backpack; 1970s; brass button; split chest tag. $150

Musical Animals: Plush animals produced in the 1980s; hang loops in heads; pull cord in bottom to activate music; plush ranging in sizes from 6.50 to 9.50 inches. Names on chest tags include Musik Peggy, Musik Piggy, Musik Pummy, Musik Lari and Gallo. Brass buttons; chest tags. Kathy Eschborn collection. $35-$40 each

"Reeves" Teddies: 3.50 and 5 inches. Two teddies secured to a wooden base with silver plaque; encased in a cello presentation box; given to Reeves International Inc. personnel on an incentive tour to the Steiff factory in 1983. Reeves was distributor of the company's products prior to the formation of Steiff U S A. Limited to 40 pieces; 1983; brass buttons; split chest tags. $1,000

Alice And Her Friends: Alice In Wonderland characters. Doll by Suzanne Gibson; 1986; Boxed. $350-$400

Teddy Babies: 9 inches. Assembled for the Toy Store's (of Toledo, Ohio) trip to Germany in 1994. Ribbons and stock tags signed by Jörg Juninger. Only 12 exist world wide. $1,000

Prototype Berryman Bear: 13 inches. The first design made under the supervision of Jörg Juninger. Eyes were not according to specifications; the tan face was to be changed to white and all parts made smaller. 1987.

Above: Prototype Berryman Bear: 13 inches. The second design; eyes still not correct as Steiff had difficulty locating side glancing style, so these pupils were hand painted; body parts to be made thinner. 1987.

Right: Prototype Berryman Bear: 9 inches. A small version that was never put into production. 1987. Final version elsewhere.

CHAPTER XX
STUDIO & DISPLAY PIECES

Large, life-size animals are important tools used by shops to attract customers to their Steiff products. Eventually they are sold and find their way into a collection. Because the number produced is limited, it is the fortunate person who is able to locate one and, in some instances, have the room for such an impressive display. The wonderful mechanical villages have been an amazing way to call attention to Steiff's booth at Toy Fairs for many, many years. The Steiff Festival, held every year since 1986 in Toledo, Ohio, has had many different mechanical exhibits at each event.

Studio Baby Boar: 22 inches. Mohair; glass eyes; open mouth; circa 1950; raised button. $1,200

Studio Fox: 22 inches. Mohair; glass eyes; open felt mouth and tongue; plastic teeth; 1950s. $1,500

Studio Tiger: 42 inches (plus tail). Unusual conformation; airbrushed striped mohair; large wooden painted eyes; metal armature in legs; swivel head; circa 1950; raised script button. $1,700

Studio Snake: 96 inches. Shaded and airbrushed mohair; open felt mouth with tongue and fangs; glass eyes; 1950s. No I. D. $5,000 up

Studio Lion: Gold mohair; long tipped mohair mane; glass eyes; open felt mouth with wooden teeth; circa 1950; raised script button. $1,250-$1,500

Above: Studio Monkey: 72 inches. Mohair; glass eyes; jointed by metal pipes; circa 1950. $3,000

Left: Studio Beaver: 34 inches. White and brown tipped mohair; glass eyes; open mouth with teeth; felt feet; swivel head; circa 1950; raised button. Hat and scarf added. Evelyn and Mort Wood collection. $1,400 up

Studio *Wittie* Owl: 24 inches. Mohair; large glass eyes; rubber beak and talons; feathers on head; swivel head; all I. D. 1950s, including hand lettered stock tag. $1,700 up

Studio Goose: 24 inches. Mohair; glass eyes; felt beak and feet; chest tag; 1950s. $1,700

Studio Dalmatian: 25 inches. White and black spotted mohair; open felt mouth; glass eyes; swivel head; leather collar; raised button; chest tag. $1,100

Studio Mountain Cock: 33 inches. Mohair; glass eyes set in sponge sockets; fighting stance; raised button; 1950s. $900 up

Studio Kingfisher: 8 inches. Plush; plastic eyes, beak and feet; mounted on a birch log; slightly soiled; circa 1975; brass button; studio chest tag. $150

Studio Puma: 29 inches. Synthetic plush; plastic eyes; circa 1975; raised script button. $900

Studio Emu: 63 inches. Mohair; velvet; felt; blue glass eyes; circa 1960; bear head chest tag. $2,000

Studio Collie: 48 inches. Long and short mohair; glass eyes; open felt mouth and tongue; circa 1950; large chest tag. $3,500

Studio Hen: 21 inches; plush; plastic eyes; vinyl beak; velour comb and feet; raised button; split chest tag; circa 1972. $500

Display *Lulac* Rabbit: 26 inches (plus ears). Gray and white plush; open velour mouth and foot pads; plastic eyes; swivel head; 1972; incised button; split chest tag. $300

Studio *Paddy* Puffin: 10 inches. Plush; felt beak, tail and feet; plastic eyes; 1980s; brass button; split chest tag; hang tag. $250

Display Teddy Bear: 30 inches. Tan mohair with clipped area around muzzle and eyes; plastic eyes; velour pads; all jointed; 1980s product. Brass button; split chest tag. $1,200

Studio Woodpecker: 11 inches. Acrylic and cotton blend; stiffened tail and wings; plastic eyes and beak; feet attached to birch log; 1980-1982; brass button; red studio chest tag. $275

Studio Woodpecker: 11 inches. Acrylic and cotton blend; stiffened wings and tail; plastic eyes and beak; attached to a birch log; 1980-1982; brass button; red studio chest tag. $275

Display *Goldy* Hamster: 20 inches. Plush; velour mouth and inner ears; plastic eyes; 1980s; brass button; split chest tag. $375 up
Display *Joggi* Hedgehog: 20 inches. Plush; velour open mouth; 1980s; brass button; split chest tag. $375 up

Studio Owl: 12 inches. Dralon; plastic eyes and beak; 1982; brass button; yellow stock tag; red paper studio tag. $300

Cosy Nosy Rhino Display: 28 inches. Plush; plastic eyes; 1980s; brass button; split tag; hang tag. $350-$375

Display *Cosy Froggy*: 24 inches. Plush; plastic eye; 1980s; brass button; split chest tag. $475 up

Display Tree House: 26 inches. Cylindrical wood with simulated bark exterior; 3 curtained windows; 4 perches; mounted on base; 1950s. No I. D. $1,000 up

Interior of Tree House: Fireman's pole; shown with bears and animals of the period.

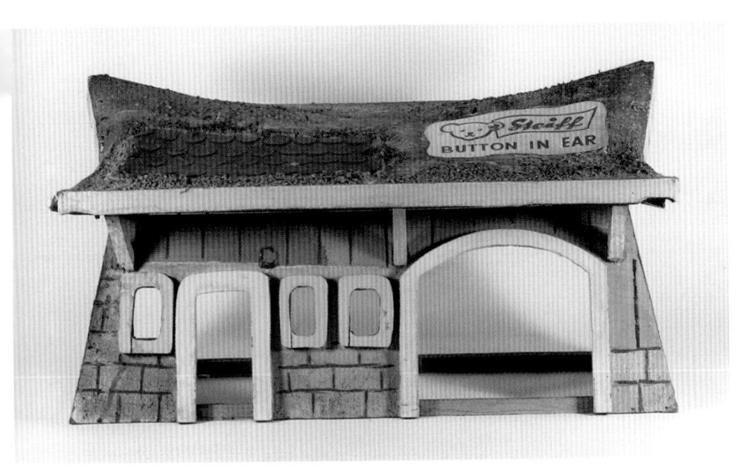

Display Farmhouse and Barn: 27 inches. Wooden; plastic and canvas overlay; 1950s; paper label on roof. $1,000 up

Wooden and plastic Ark: 30 inches (approximately). Palm trees missing; shown with animals of the period; no I. D. 1950s. $1,000 up

Mechanical Village: Very large village in old world style; animals and bears do a variety of animated movements; circa 1960.

Another view of the mechanical village.

View of mechanical village.

View of mechanical village.

View of mechanical village.

Wooden display rack: 24 inches (high). Plywood with pine head; two bins to display toys; 1980s; large brass button with cardboard stock tag. $150
Wooden peg rack: 19 inches long. Pine with five peg hooks for hanging objects; large brass button with cardboard stock tag. $150

Another Mechanical Village on display at the Steiff Festival in Toledo, Ohio.

Mechanical Village House on display at the Steiff Festival in Toledo, Ohio.

CHAPTER XXI
FUN & FANCIES

Postcards, advertising stamps and limited editions to herald a celebration honoring a special group are among the fun things to add to a collection. Books and products, not made by Steiff, but incorporating their wares are also items that are highly sought.

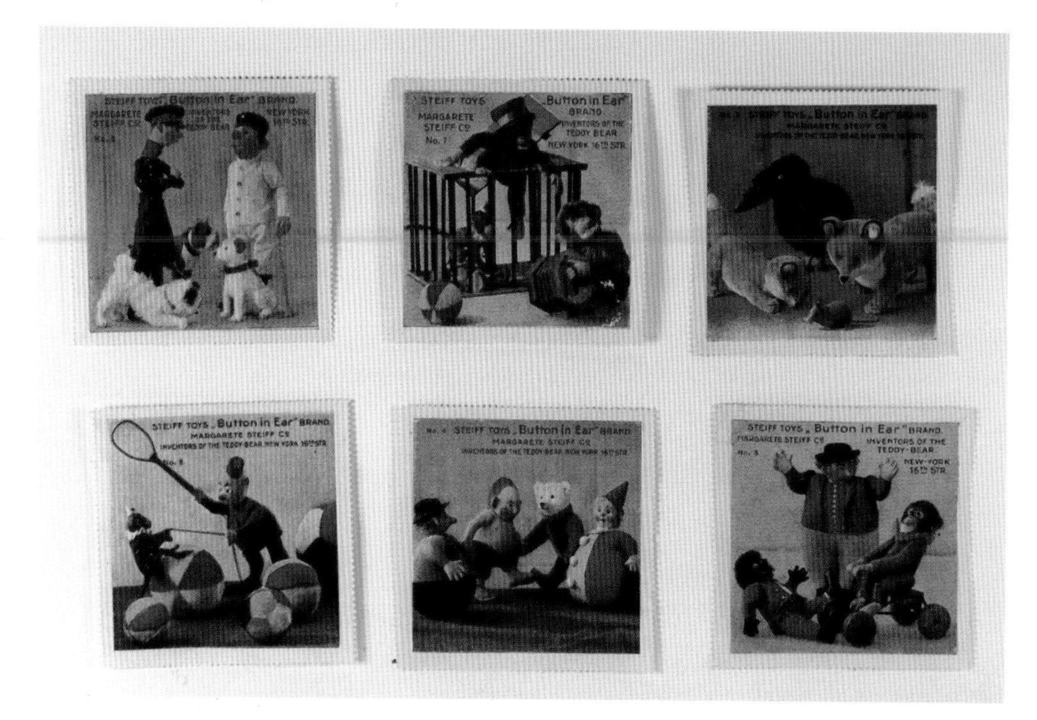

Stamps: Advertising stamps showing several Steiff products. Legend printed on each stamp. $200 each

Circus Cage: 11 inches. Colorful wagon; top lifts off; rubber tired metal small wheels; front axle turns; 1957; logo painted on end. $450-$475

Circus wagon: 11 inches. Colorful wagon; top lifts off; rubber tired metal wheels; 1957; logo painted on end. $450-$475

Blocks: colorful wooden blocks; bear head logo and Hohlkubus printed on one side of each block; circa 1925; paper label on box with logo and no. 805. $450

Hobby Horse: 40 inches. Tan and white felt horse head; bristle mane; brown glass eyes; bridle, bit and reins; wooden pole and wheels; raised button. $395

Brochure: printed booklet showing the horse and Roosevelt figure; available in both English and German languages; 1950s; originally came with the horse. $25

Cage: 9 inches. Wooden cage that originally was part of a set; dowel missing; circa 1950; "Steiff Button In Ear" painted on top. Shown with Lion of the period. $125-$135

Kite: Model called *Adler*; shaped like a Falcon; complete with carrying case; circa 1950; raised script button on both kite and carrying case; logos printed on case. $450

Max and *Moritz*: 4 inches; plastic and felt figures; shown with German book that featured them; all I. D. 1960s. $250 for all

City Mouse House: Assembled by F. A. O. Schwarz in the 1960s; furniture may vary; came with Pieps mouse dressed in tulle; F. A. O. Schwarz sticker. $350

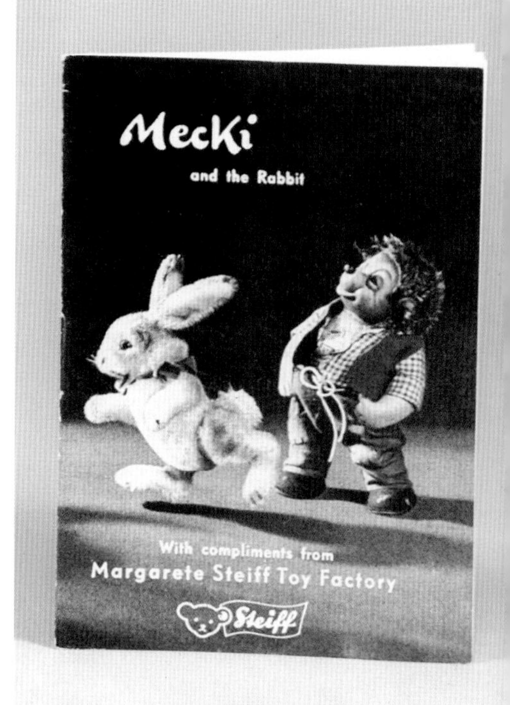

Above: *Mecki* booklet: 4 inches. Booklet printed in Germany that tells the story of the *Mecki* and *Micki* Hedgehog family. C. Diehl Brothers. Also shown on cover is *Niki* rabbit; circa 1958. $25

Left: Playing Cards: each card features an animal; in plastic case; 1960s. $250

French Opel car: Logo and Bully dog applied to car; made in France for Canadian market. $125

Niki Rabbit Postcards: four views of circa 1950s rabbits; printed in Germany. Nach Steiff Originalen. $10-$15 each

Postcards: circa 1960 postcards featuring *Zotty* bears; logo and address on back. $10 each

Above: Book: Teddy Und Seine Freunde. A Kinderbuck featuring many Steiff animals; printed in 1975. $65

Left: *Vincent*, 2 Lambs and *Klaff Dog*: 4 and 7 inches. Vinyl and felt sheep herder; plush.
Snucki and *Flori* Lambs: plush dog; rare to find set complete and in cello box; 1970s; incised buttons; split chest tag; hang tag. $500

Roly Poly Clown: 10.50 inches. Mohair and felt; replica of 1909; button and hang tag; 1988. $325

Hankies: Two cotton hankies featuring a bear and a cat; one of several items produced in the 1980s. $30-$35

Bear Head Pin: 1.50 inches. Mohair head with red bow; came originally in four colors (honey, caramel, white and brown); 1985; brass button. $65

Above: Towel: Terrycloth with appliquéd bear; in presentation box; 1980s. $10-$15

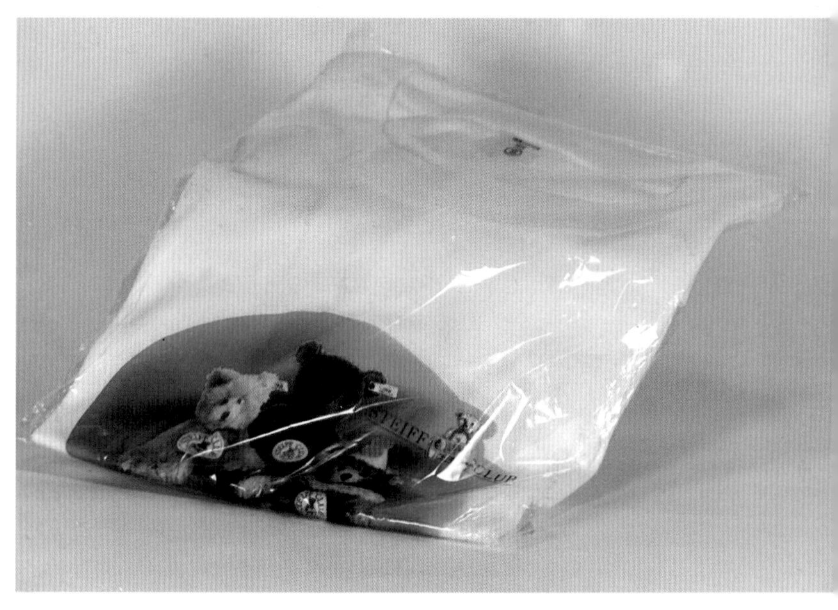

Right: Tee Shirt; 1990s. $18

Advertising Cat: 6 inches. Plush advertising *Whiskas* cat food; button; chest tag; hang tag. 1990s. $125

Porcelain Cat: 4 inches; 1995. $65

Porcelain Bear Brooch: 2 inches; boxed. 1995. $45

Porcelain Elephant Brooch: 2 inches; boxed. 1995. $45

Porcelain Cat Brooch: 2 inches; boxed. 1995. $45

Festival of Steiff Plate: Souvenir for convention guests. 1990. $25-$40

Festival of Steiff Mug: Souvenir for convention guests. 1993. $25-$40

Postcard replica of 1908 punting party scene. 1990s. $2-$5

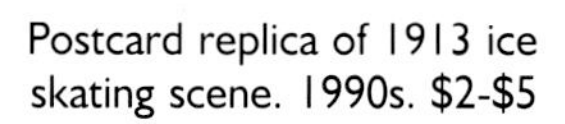
Postcard replica of 1913 ice skating scene. 1990s. $2-$5

German Paper Plate: 10 inches. Colorful Christmas paper plate adorned with snow scene featuring Steiff animals and birds; circa 1955. $55

German Paper Plate: 12 inches. Village scene that features a Steiff Dalmatian; circa 1950s. Made in Western Germany. $30-$35

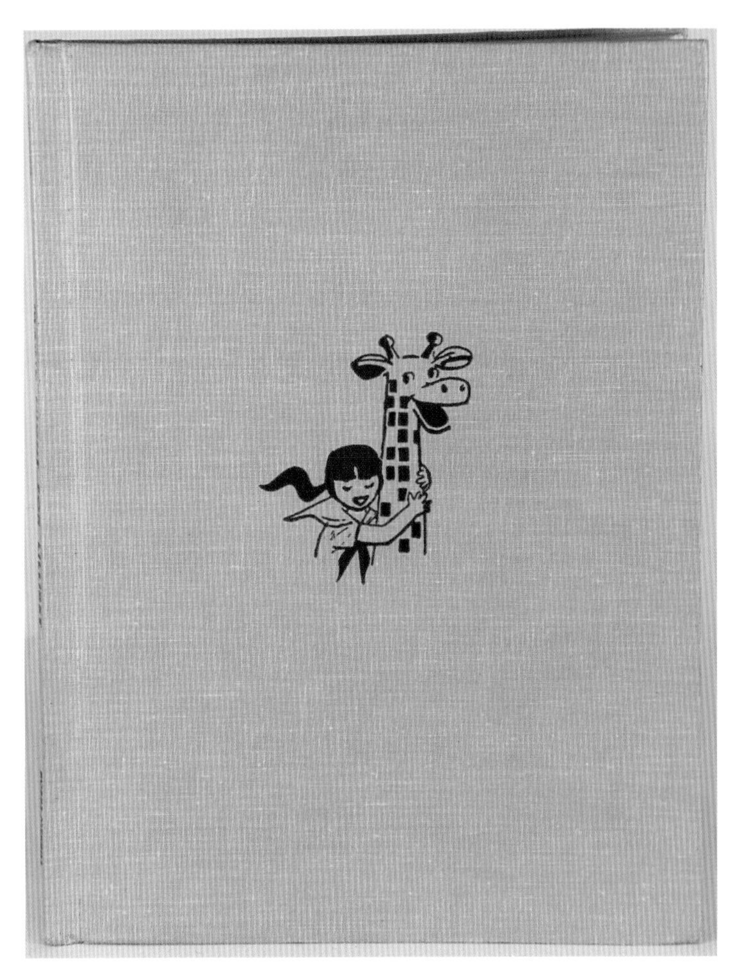

Book:*Magic Night For Lillibet*. Story and photographs by Gerry Turner; drawings by Ralph Owen; Bobbs-Merrill Company; 1959. The story of a child and her giraffe who was said to have run away; the bulk of the book takes place at Lazarus Dept. Store in Columbus, Ohio. It was printed to showcase a large giraffe on display and various other Steiff animals, including studio pieces. $65-$75

Book: *The Perfect Zoo*. By Eleanor Farjeon; features all Steiff animals; 1929. Published by George Harrap Co., London. $450

CHAPTER XXII

CATALOGS & IDENTIFICATION MARKS

It goes without saying that before embarking on the collection of *anything,* one should arm oneself with as much knowledge as possible. It is particularly true of Steiff toys since so many items are falsely touted as Steiff and buttons have been placed in the ears of non-Steiff toys as well.

Original catalogs and reproductions of catalogs, often available for ten dollars and up, are a wise investment. The studying of identification marks should be undertaken, bearing in mind that this can be a tricky dating method. Often the company used older buttons on newer products if that was all that was available. This is not a problem with advanced collectors, but the novice would do well to consult a reliable dealer or a knowledgeable friend with the same interest.

The following is a dictionary of marks and the dates they were in use.

Buttons attached by two prongs:

1904-1905. Silver metal with embossed elephant.

1904-1905. Silver metal that is blank.

1905-1930s. Silver metal with Steiff printed and embossed. The final F underscores the first F as FF.

1948-1950. Slate blue painted blank button.

1948-1950. Blank blue button.

1950. Steiff embossed in capital letters.

1952-mid-1960s. Silver button with Steiff in raised script lettering.

Buttons attached by rivets:

1967-1977. Silver metal with incised script lettering.

1977 on. Brass button with Steiff in incised script.

Stock tags used behind the button:

Circa 1908. White with four numbers and the word "geschützt."

1908-1911. White with four numbers and the words "Original Steiff."

1911-1926. White label with about five numbers and the words "geschützt Germany imported Allemagne."

1926-1935. Red / orange label with the words "Steiff original geschützt Made in Germany" and five numbers.

1935-1943. Yellow label with the same words as the red / orange one. The label remains yellow up to the present with the exception of limited editions. They revert to the white color, although the material changed from woven cloth to a folded ribbon.

Paper labels. Usually attached to the chest. On dogs it is normally affixed to the collar and on doll figures it is attached to a plastic bracelet. The generic term is to call it a "chest tag" regardless of where it is placed.

1900-1904. Round red and white paper label with an elephant in the center and printing around the edge.

Labels do not appear for several years between 1904 and 1926.

1926-1928. Metal rimmed white paper circle with the animal's name printed in bold and "Steiff original" printed below.

1928-1950. White circle with a red edge and a square headed yellow teddy at the bottom. The words are the animal's name in the center with "Steiff-original-marke" around the edge.

1950-1972. The same as number four except the teddy head is rounded.

1972 on. Known as a split tag. Divided in half with color yellow at top and red on the bottom. The animal's name appears in the yellow half and "Steiff Knopf im ohr" on the bottom.

Replicas have a variety of reproduction stock tags and chest tags.

Additionally, a white fabric seam tag will appear on animals from 1946-1953 that reads made in U S Zone Germany. Sometimes only a remnant of the tag remains.

Approximately ninety-eight percent of the products illustrated in this book have been handled and sold by me during the course of my years in business. Therefore, the prices are as accurate as possible. However, please keep in mind that there are variables that are reflected by condition, identification marks and rarity. You may find a toy priced higher or lower. I always consider a price guide helpful in considering whether the value is closer to $100 or $1,000. That is why it is called a "guide." Nothing is written in stone. In the final analysis, a collector should pay what he or she is comfortable with. Neither this author nor the publisher is responsible for any losses incurred when either buying or selling. My years of buying, selling, and collecting Steiff have been joyful. What a thrill it is to finally find the elusive piece you have been longing for. I wish you all "Happy Hunting."

The Steiff Company produces a product catalog every year. The following pages contain a selection of what has been available over a span of time.

1999-2000 "Exclusively Yours" catalog

Exclusively
Yours
Steiff
BUTTON IN EAR
The World's Oldest
Plush Company
www.steiffusa.com

Help Keep Him Safe
This mohair koala, clutching a bunch
of eucalyptus leaves, was created by Steiff
in partnership with the World-Famous
San Diego Zoo® and their Center for
Reproduction of Endangered Species (CRES).®
Steiff is proud to donate a portion of the
proceeds from the sale of this koala to benefit
CRES in their efforts to find new ways to
help reproduce rare animals that can no longer
survive on their own in the wild.
San Diego Zoo Koala
EAN #665967 12"
North American Limited Edition of 1,500
$325
16

Gulliver Travels
Meet Gulliver, looking cheerful in his bright
red sweatshirt. He's Steiff's interpretation of
the "Spokesbear" for Good Bears of the World.
The original Gulliver was a traveling
ambassador for GBW, a charitable organization
that uses teddy bears to comfort children in
times of trauma. Your purchase of Steiff's
Gulliver helps give these kids friendly teddies
to hug — a portion of the proceeds goes
directly to GBW for their charitable work.
Gulliver
EAN #665936 16"
North American Limited Edition of 5,000
$325
1

Fresh as the Morning Dew
The uncommonly crimson Dew Drop Rose,
second in Steiff's series of richly colored
designer bears, looks as if she's just
returned from a walk in the garden.
Clinging to her face is an Austrian Crystal
dew drop, and she holds a velvet rose.
Her coat is luxurious, custom-dyed mohair
and she pads around with paws of black suede.
Dew Drop Rose
EAN #665844 16"
North American Limited Edition of 3,500
$325
2

Millennium Bear
A brand new bear to greet
the new century with Steiff style.
Millennium Bear features five-way jointing,
a growler, rich mohair fur, and classic charm.
He shows the world that Steiff
intends to take its timeless tradition
of quality right into the
next millennium.
Millennium Bear
EAN #670374 16"
Worldwide Limited Edition – Open until the end of 2000
$360
3

Tee-Off Time

You may know him as Jack Nicklaus, Golfer of the Century, but he's affectionately known as "The Golden Bear," and Steiff introduces this special teddy in his honor. The golden mohair bear features Jack's signature embroidered on the paw-- and comes with three custom Maxfli golf balls! This is first in a series of bears dedicated to legendary sports figures.

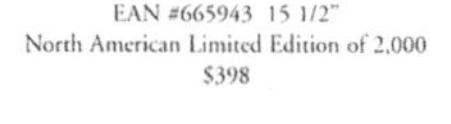

Jack Nicklaus Bear
EAN #665943 15 1/2"
North American Limited Edition of 2,000
$398

4

COCA-COLA Polar Bear

Meet the most refreshing Steiff bear ever created! In an unprecedented collaboration with The Coca-Cola Company, Steiff presents a frosty polar bear ready to take a sip from an ice-cold bottle of the world's favorite beverage. He keeps away the chill with a knitted COKE Red scarf and wears an enamel medallion.

COCA-COLA Polar Bear
EAN #670336 15 1/2"
Worldwide Limited Edition of 10,000
$375

5

A Friend from the 100-Acre Wood

He sports a familiar red vest and a famous round tummy. Try as he might, he can never get quite enough honey. He's none other than Winnie the Pooh, and he's the first in a charming Steiff series based on the characters of A.A. Milne. This beloved bear will step out of the 100-Acre Wood and right into your heart.

Winnie the Pooh
EAN #651489 11"
Worldwide Limited Edition of 10,000
$250

6

Set Sail with Mickey

Steamboat Willie, with his handcrafted ship's wheel, is a plush portrayal of Mickey Mouse in his first-ever animated movie performance. Part of the Disney Showcase Collection, this mohair and velvet piece honors a longstanding Steiff-Disney tradition that began back in the 1930s.

Steamboat Willie
EAN #651472 14"
Worldwide Limited Edition of 10,000
$375

7

Message of Peace

What better wish for a new millennium than peace throughout the world? This gentle Steiff bear, wearing a white dove medallion around his neck, proclaims our hope for a new era of cooperation and love among nations.

Peace Bear
EAN #665981 11 1/2"
North American Limited Edition of 3,000
$198

8

Honor the Old Welcome the New

Join Steiff in celebrating a new century, while bidding the old one farewell. The blue tones of Goodbye 1999 deepen as he fades off into the night. Hello 2000 goes from pale yellow to vibrant rose as a new era dawns. The bears arrive as a set in a commemorative, satin-lined presentation book.

Hello Goodbye Set
EAN #670367 Each bear 8 3/4"
Worldwide Limited Edition – open until the end of 2000
$450

9

The Wind in the Willows Collection

Ratty is positively dapper in a casual white sailing outfit. Badger looks smart in a snappy three-piece suit. The classic children's tale continues with two new Steiff friends, fresh from the riverbank, carefully based on the drawings of E. H. Shepard.

Ratty
EAN #037047 10"
Worldwide Limited Edition of 4,000
$260

Badger
EAN #037039 11"
Worldwide Limited Edition of 4,000
$325

10

Homespun Hero

(Book not included)

If you imagine skipping school and rafting down the Mississippi, one freewheeling character comes to mind — Huckleberry Finn. Now this all-American hero is brought to life as a Steiff bear complete with patched overalls, a real wooden pipe and a mischievous twinkle in his eye. He'll be joined next year by his pal and partner in crime, Tom Sawyer.

Huckleberry Finn Bear
EAN #665851 11 1/2"
North American Edition of 1,500
$275

11

A Teddie to Cherish

Cherished Teddies are adored by figurine collectors around the world for their messages of love. We are proud to introduce Daisy, a mohair bear based on the classic Cherished Teddies figurine titled "Friendship Blossoms with Love" which was retired in 1996. This is the first in a series of three Steiff Cherished Teddies being created in collaboration with Enesco.

Enesco Daisy
EAN #665981 13"
North American Limited Edition of 5,000
$295

12

Party Animal

Blow up the balloons, bake the cake, wrap the presents, light the candles. With a red satin birthday hat, a festive ribbon, and a birthday message embroidered on his paw, this new Steiff bear is ready to celebrate. To add to the fun, he's musical! Wind him up and listen to a charming rendition of "Happy Birthday to You."

Birthday Bear
EAN #665547 10"
North American Open Edition
$225

13

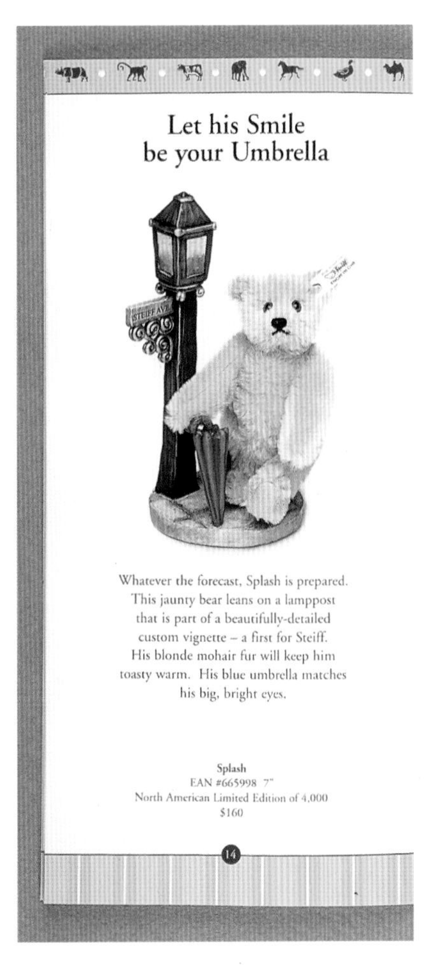

Let his Smile be your Umbrella

Whatever the forecast, Splash is prepared. This jaunty bear leans on a lamppost that is part of a beautifully-detailed custom vignette – a first for Steiff. His blonde mohair fur will keep him toasty warm. His blue umbrella matches his big, bright eyes.

Splash
EAN #665998 7"
North American Limited Edition of 4,000
$160

14

Deck the Halls

Steiff sends good cheer your way with two new holiday treasures. Noel, wearing Christmas colors and an advent wreath headdress, is the first in a series of Steiff Christmas bears. All Wrapped Up is a diminutive bear poking out of a tiny gift package, sixth in a series of Steiff ornaments. Add them both to your Christmas list this year!

All Wrapped Up
EAN #665882 4"
North American Limited Edition of 5,000 $110
Noel
EAN #666025 15 1/2"
North American Limited Edition of 1,500 $250

15

1999.

1990. Reprint of 1928 Catalog.

1994.

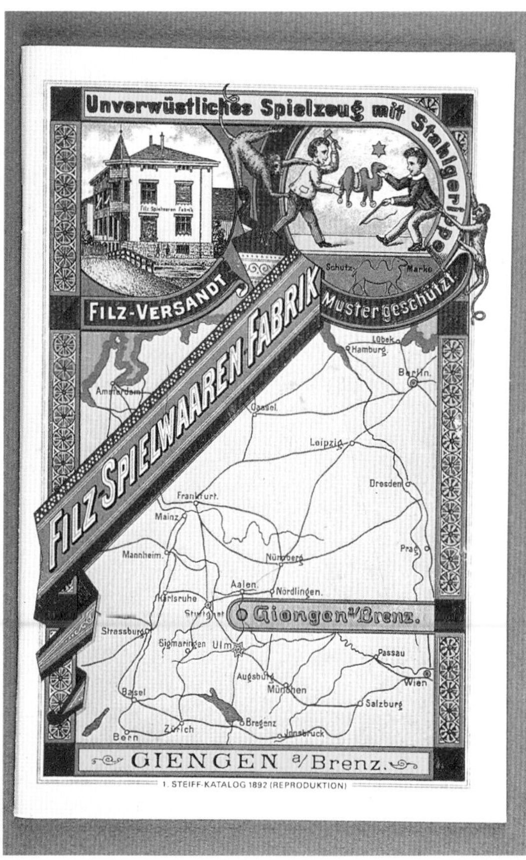

Modern reprint of Steiff's first catalog - 1892.

FILZ-VERSANDT-GESCHÄFT GIENGEN A. D. BRENZ.

Kameele.

No. 1.

Reit-Kameel.

56 cent. hoch, auf Räder Mark 16.— bis Mark 20.—.
Stahlgerippe aus Mannesmannröhren, durchaus unverwüstlich, für Knaben bis zu 5 Jahr.
Auf Wunsch wird dieses Modell noch grösser mit Wiege angefertigt.

Modern reprint of Steiff's first catalog - 1892.

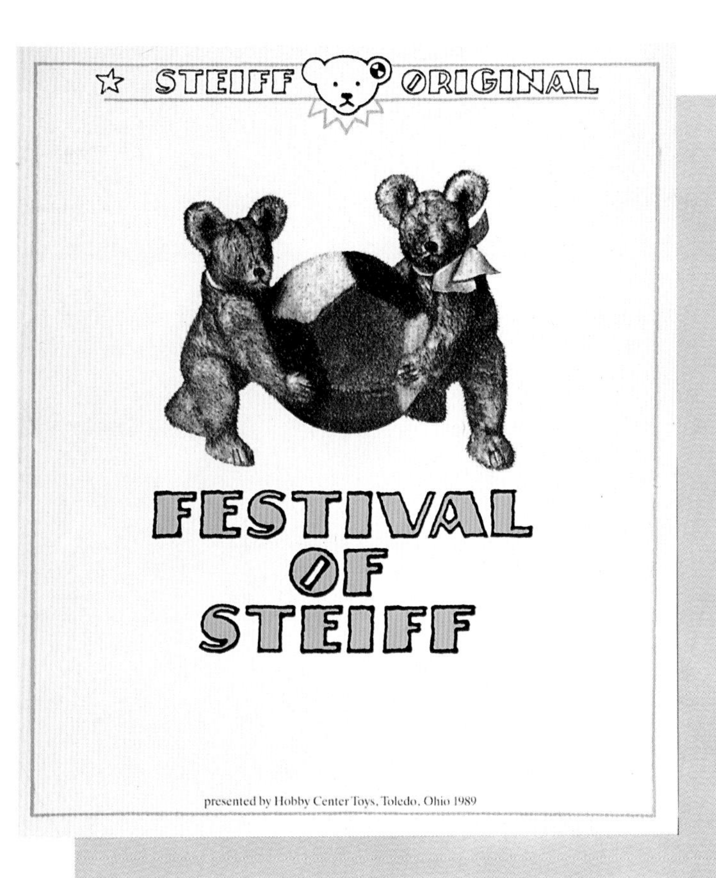

STEIFF ORIGINAL

FESTIVAL OF STEIFF

presented by Hobby Center Toys, Toledo, Ohio 1989

STEIFF ORIGINAL

12d

Petsy	Petsy	Petsy
Der junge Teddy	The young Teddy Bear	El Teddy joven
gegliedert, langh. Mohairplüsch braun gespitzt, weich, rote Nase, blaue Augen, Seidenschleife, Brummst.	jointed, long hair Fine Plush brownish mingled, soft, red nose, blue eyes, silk ribbon, voice	articulado, felpa mohair pelo largo marron, blando, nariz roja, ojos azules, lazo de seda, voz

kg	No.	stehend cm	standing ins
0.10	5315	22	9
0.15	5317,2	25	10
0.24	5320,2	30	12
0.28	5322,2	32	13
0.34	5325,2	35	14
0.60	5328,2	42	17
0.73	5332,2	46	18
0.89	5335,2	53	21
1.53	5343,2	65	26
2.—	5350,2	75	30

11 Bully

stehend ohne Räder, Mohairplüsch schwarz oder braun | standing without wheels, Fine Plush black or brown | deredo sin ruedas, felpa mohair marron ó negra

kg	No. (cm)	height inches
0.07	1314,0	5½
0.13	1317,0	7
0.19	1322,02	9
0.34	1328,02	11
0.56	1335,02	14

17 Charly

stehend, langhaar. Mohairplüsch hellbraun/weiß, drehbarer Kopf | standing, long hair Fine Plush lightbrown/white, turnable head | deredo, felpa mohair pelo largo, rubio/blanca, cabeza movible

kg	No. (cm)	height inches
0.02	1307,0	3
0.04	1310,0	4
0.06	1312,0	5
0.08	1314,0	5½
0.12	1317,0	7
0.23	1322,0	9
0.40	1325,02	10

17 Charly

sitzend, langhaarig Mohairpl. hellbraun weiß, drehbarer Kopf

sitting, long hair Fine Plush lightbrown white turnable head, voice

sentado, felpa mohair pelo largo rubio blanca, cabeza movible, voz

kg	No. (cm)	height inches
0.04	3310	4
0.08	3314	5½
0.13	3317	7
0.25	3322,2	9
0.34	3325,2	10
0.43	3330,2	12
0.77	3335,2	14

38

Jagdhund (Treff)
sitzend, Mohairplüsch hellbraun drehbarer Kopf, Stimme

Pointer
sitting, Fine Plush lightbrown, turnable head, voice

Perro de Caza
sentado, felpa mohair rubia, cabeza movible, voz

	kg	No.	height inches
Treff	0.05	3314	5½
	0.085	3317	7
	0.180	3322	9
	0.26	3328	11
	0.51	3335	14

35

Kücke	Chicken	Pollito
stehend, Mohairplüsch gelb braun	standing, Fine Plush yellow brown	deredio, felpa mohair amarilla marron

kg	No.	height inches
0.014	1310	4
0.034	1314	5½
0.053	1317	7

36

Hase	Rabbit	Conejo
liegend, Mohairpl. weiß, drehbarer Kopf	lying, Fine Plush white, turnable head	acostado, felpa mohair blanca, cabeza movible

kg	No.	height inches
0.017	2305	2
0.025	2306	2½
0.034	2307	3
0.045	2309	3½
0.075	2311	4½
0.125	2313	5
0.165	2315	6
0.290	2318	7

STEIFF ORIGINAL

44

Katz	Tabby-Cat	Gato
stehend, Mohairplüsch, hellbraun gestromelt	standing, Fine Plush lightbrown striped	deredo, felpa mohair rubio listado

	kg	No. (cm)	height inches
Tabby	0.018	1307,0	3
	0.034	1310,0	4
	0.05	1312,0	5
	0.06	1314,0	5½
	0.11	1317,0	7
	0.17	1322,0	8

dasselbe auf Exz.-Rädern / on eccentric wheels / sobre ruedas excentricas

kg	No.	height inches
0.17	1314ex	5½
0.23	1317ex	7
0.36	1322ex	9

liegend, Mohairplüsch hellbraun gestromelt | lying, Fine Plush lightbrown striped | acostado, felpa mohair rubio listado

	kg	No.	height inches
Tabby	0.018	2305	2
	0.026	2306	2½
	0.037	2308	3
	0.065	2310	4
	0.10	2312	5
	0.15	2314	5½
	0.22	2317	7
	0.31	2320	8

54

Pinguin	Penguin	Penguino
stehend, Mohairplüsch schwarz/weiß	standing, Fine Plush black/white	deredo, felpa mohair blanco/negra

kg	No.	height inches
0.05	1317,0	7
0.10	1322,2	9
0.18	1328,2	11
0.30	1335,2	14

74

Ziege	Goat	Cabra
stehend, Mohairplüsch weiß, Seidenschleife	standing, Fine Plush blanco, silk ribbon	deredo, felpa mohair blanca, lazo de seda

kg	No.	height inches
0.07	1317,0	7
0.13	1322,0	9
0.22	1328,0	11

85

Tier-Handtaschen	Animal hand bags	Bolsas de mano
Mohairplüsch mit Reißverschluß	Fine Plush with hookless fastener	felpa mohair

kg		No. (cm)	height inches
0.11	Schimp. 9	5317P	7
0.09	Bear 12	1317P	7
0.17	Bully 11	1322P	9
0.22	„	1328P	11
0.23	„	3325P	10
0.19	Charly 17	1322P	9
0.29	„	1325P	10
0.10	Molly 68	1317P	7
0.19	„	1322P	9
0.24	„	1325P	10
0.34	„	1328P	11
0.10	„	3317P	7
0.26	„	3325P	10

88

Karikaturen	Comic dolls	Muñecas caricaturas
Clown aus hellblau Samt	Clown with suit of lightblue velvet	Clown de terciopelo azul

kg	No.	height inches
0.12	428	11
0.21	435	14
0.34	434	17

97

Music Tiere	Musical Toys	Animales con música
Tiere aus Mohairplüsch, Musik durch Drücken des Körpers	Animals of Fine Plush, play by squeezing the body	Animales de felpa mohair, Música apretando el cuerpo

kg		No.	height inches
0.56	Teddy 12	5330,3	12
0.45	Charly 17	1325,3	10
0.52	Molly 68	1328,3	11
0.40	Molly	3325,3	10
0.41	Fluffy 44	3330,3	12

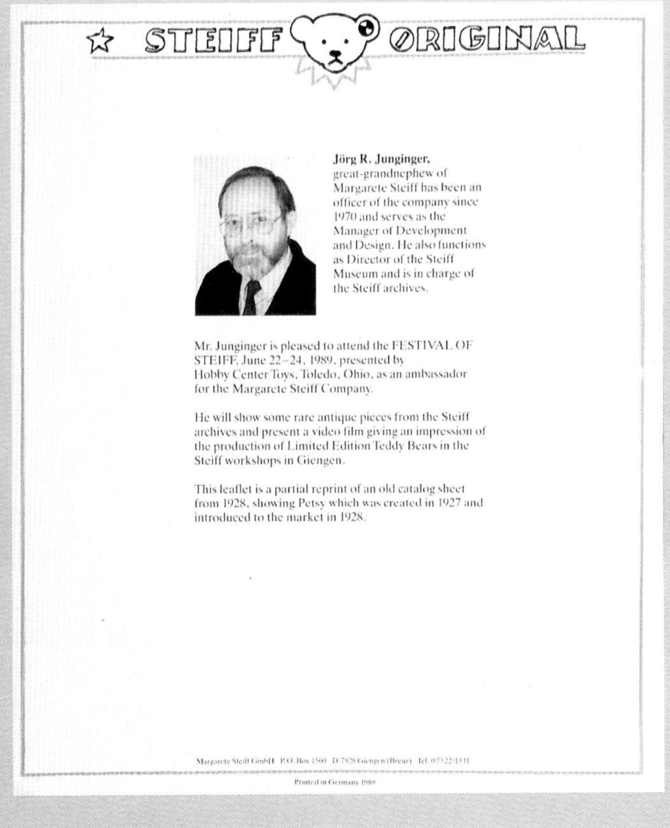

STEIFF ORIGINAL

Jörg R. Junginger, great-grandnephew of Margarete Steiff has been an officer of the company since 1970 and serves as the Manager of Development and Design. He also functions as Director of the Steiff Museum and is in charge of the Steiff archives.

Mr. Junginger is pleased to attend the FESTIVAL OF STEIFF, June 22–24, 1989, presented by Hobby Center Toys, Toledo, Ohio, as an ambassador for the Margarete Steiff Company.

He will show some rare antique pieces from the Steiff archives and present a video film giving an impression of the production of Limited Edition Teddy Bears in the Steiff workshops in Giengen.

This leaflet is a partial reprint of an old catalog sheet from 1928, showing Petsy which was created in 1927 and introduced to the market in 1928.

Margarete Steiff GmbH · P.O. Box 1560 · D-7928 Giengen (Brenz) · Tel. 07322/131

Printed in Germany 1989

1989 Reprint of a 1928 catalog.

1909: Richard Steiff mit einem Handmuster seines Teddybären

Printed in W. Germany 30.000 d 2/89

Modern reprint of a 1909 photograph of Richard Steiff.

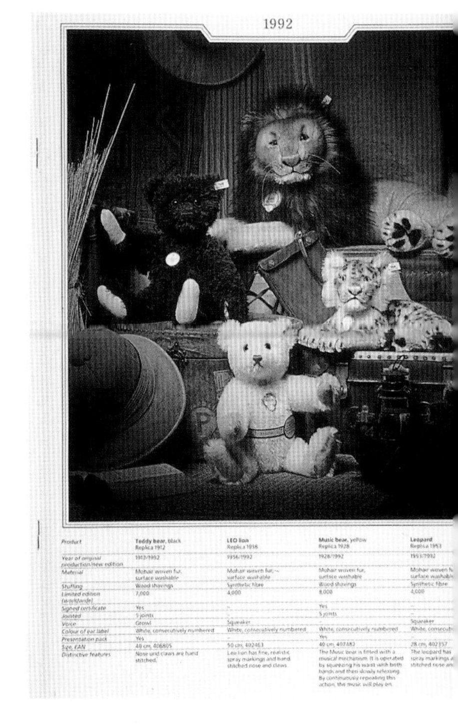

1992

1992.

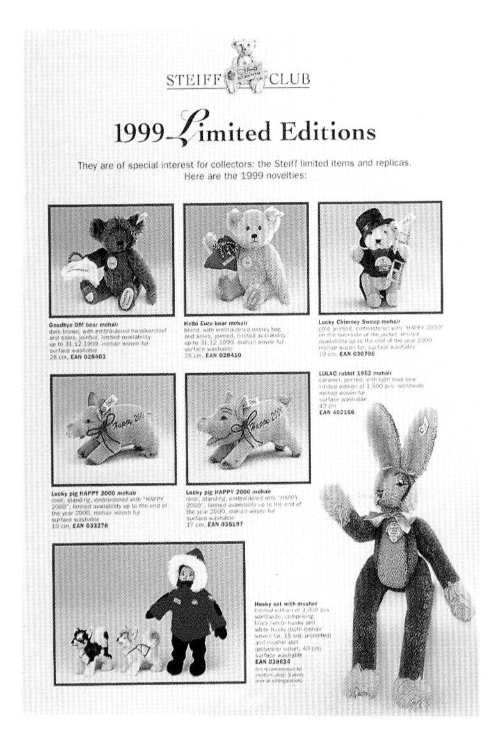

STEIFF CLUB

1999 Limited Editions

They are of special interest for collectors: the Steiff limited items and replicas.
Here are the 1999 novelties:

1999.

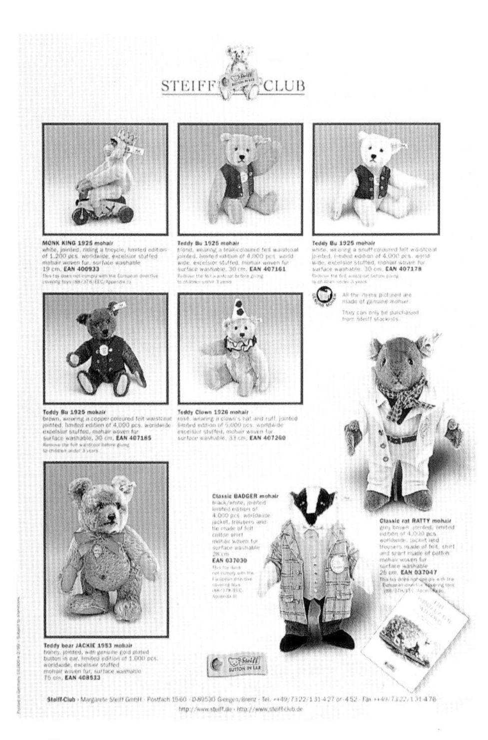

STEIFF CLUB

1999.

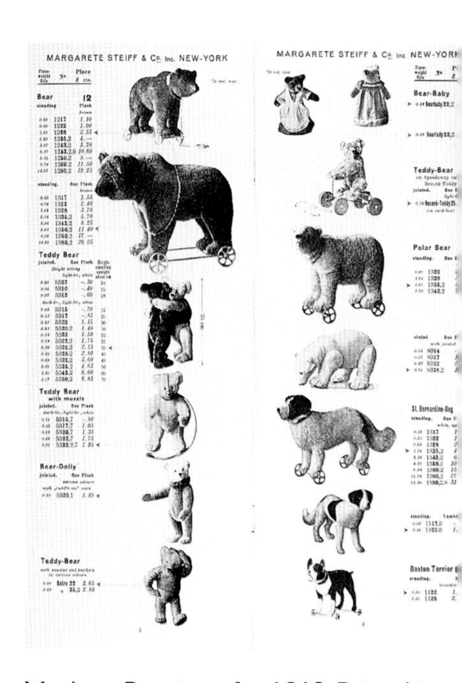

MARGARETE STEIFF & Co. Inc. NEW-YORK

Modern Reprint of a 1913 Price List.

Margarete Steiff Dolls

Natural Grace: It requires careful study and a sensitive touch to create such life-like personalities with such cheerful charm – the inimitable character of Steiff.

1989.

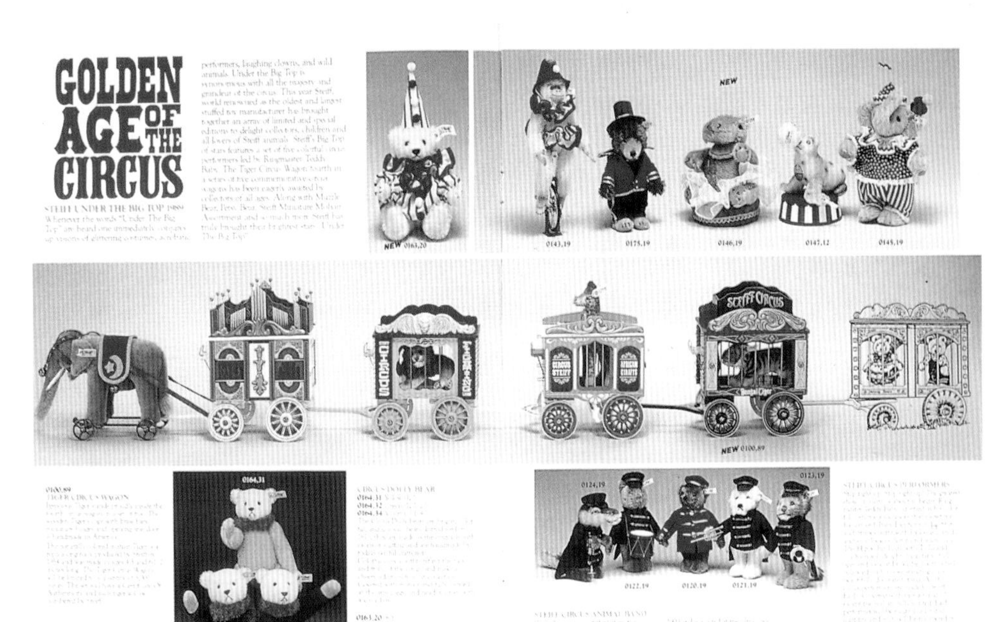

GOLDEN AGE OF THE CIRCUS
NEW
BUTTON IN EAR

1989.

Steiff
KNOPF IM OHR
BUTTON IN EAR
BOUTON À L'OREILLE

SCHAUTIERE
DISPLAY ANIMALS
ANIMAUX D'ÉTALAGE

1967.

KNOPF IM OHR

DIS 0339

Zoo-Bär/Zoo Bear/Ours zoologique
0339/12 120 cm 48"

958 DIS 10 7 67 Germany

KNOPF IM OHR

Braunbär stehend - Bear standing	0409/10	100 cm 40.00"
Braunbär aufwartend - Bear	0409/19	190 cm 75.50"
Jungbär - Young Bear	0409/08	80 cm 31.50"

BIBLIOGRAPHY

Cieslik, Jürgen and Marianne. *Button In Ear.* Germany, 1989.

Pfeiffer, Günther. *Steiff Sortiment. 1947-1995.* Germany, 1995.

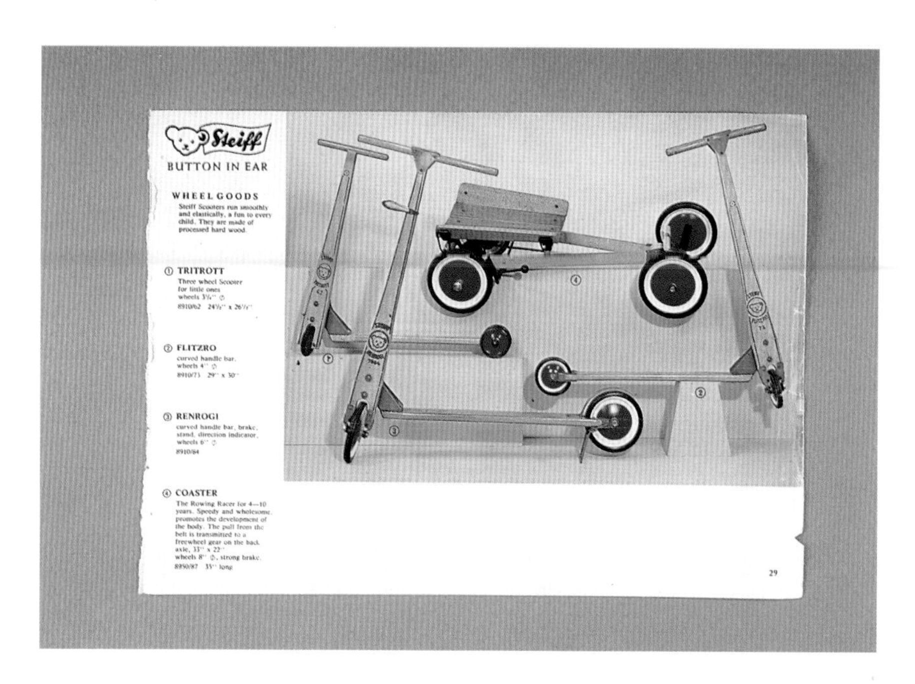

Steiff

BUTTON IN EAR

WHEEL GOODS

Steiff Scooters run smoothly and elastically, a fun to every child. They are made of processed hard wood.

① TRITROTT
Three wheel Scooter for little ones
wheels 3¼'' ⊘
8910/62 24½'' x 26½''

② FLITZRO
curved handle bar,
wheels 4'' ⊘
8910/73 29'' x 30''

③ RENROGI
curved handle bar, brake, stand, direction indicator,
wheels 6'' ⊘
8910/84

④ COASTER
The Rowing Racer for 4—10 years. Speedy and wholesome, promotes the development of the body. The pull from the belt is transmitted to a freewheel gear on the back axle, 33'' x 22''
wheels 8'' ⊘, strong brake.
8950/87 35'' long

29

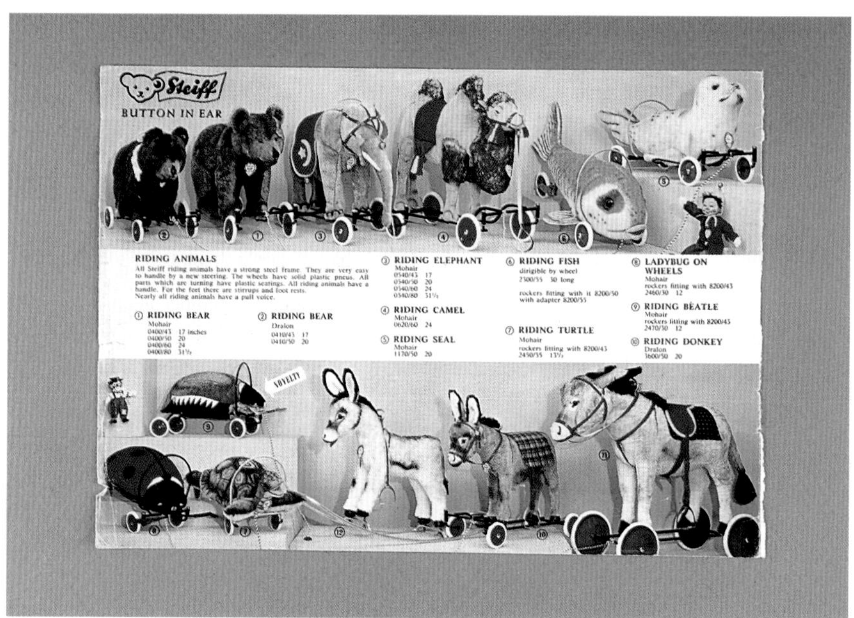

Steiff

BUTTON IN EAR

RIDING ANIMALS

All Steiff riding animals have a strong steel frame. They are very easy to handle by a new steering. The wheels have solid plastic pneus. All parts which are turning have plastic seatings. All riding animals have a handle. For the feet there are stirrups and foot rests.
Nearly all riding animals have a pull voice.

① RIDING BEAR
Mohair
0400/43 17 inches
0400/50 20
0400/60 24
0400/80 31½

② RIDING BEAR
Dralon
0410/43 17
0410/50 20

③ RIDING ELEPHANT
Mohair
0540/43 17
0540/50 20
0540/60 24
0540/80 31½

④ RIDING CAMEL
Mohair
0620/60 24

⑤ RIDING SEAL
Mohair
1170/50 20

⑥ RIDING FISH
dirigible by wheel
2300/55 30 long

rockers fitting with it 8200/50
with adapter 8200/55

⑦ RIDING TURTLE
Mohair
rockers fitting with 8200/43
2450/55 13½

⑧ LADYBUG ON WHEELS
Mohair
rockers fitting with 8200/43
2460/30 12

⑨ RIDING BEATLE
Mohair
rockers fitting with 8200/43
2470/30 12

⑩ RIDING DONKEY
Dralon
3600/50 20

NOVELTY

INDEX

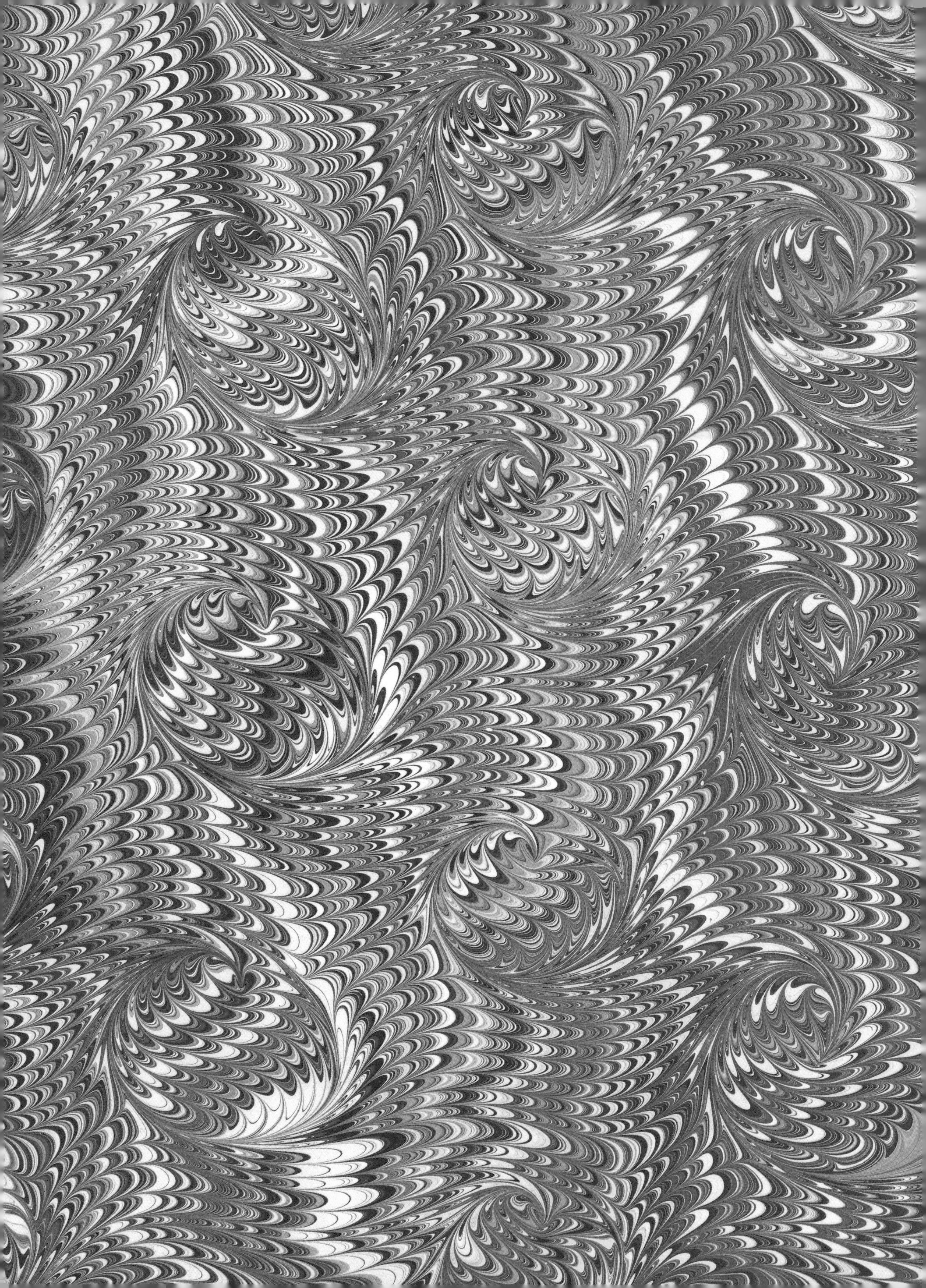